Modern Algebra

Part 1: Families of Finite Groups

by

Carroll W. Boswell

copyright 2018 Carroll W. Boswell

ISBN - 13:978-1983545054
ISBN – 10:1983545058

Preface to Modern Algebra part 1

There are so many excellent introductory algebra textbooks that some remarks should be made as to what distinguishes this one from the others, what its advantages and disadvantages are. The algebra texts I am familiar with are the ones by Gallian, by Fraleigh, by Cheryl Chute Miller, and by Dumit and Foote. They vary in how demanding they are of the student, in how much detail they cover, and in how quickly they pace the material. They all have this one thing in common: they are aimed at mathematics majors; that is, students who have some sophistication in mathematics already, who have sufficient background in college level mathematics to have a reasonable chance of success with abstract algebra. I am aiming at students who are more at risk, but who still want to do real mathematics.

I believe there are many people who are capable of abstract thought on this level but who lack the background or the facility with abstraction to keep up with the pace of the typical mathematics textbook. These are students who have an interest in learning some of the "hard" math but who need considerably more in the way of concrete examples and routine problems, more practice doing grungy calculations, and who need a slower pace. These are the students I have in mind. I am hoping this book makes mathematics accessible to a much wider audience than it is currently. The proficient math major will find the pace of this book too plodding, too filled with obvious details, etc. Such students will be perfectly happy with one of the many other excellent algebra texts I mentioned in the first paragraph.

I believe that certain portions of this book will be quite challenging to the students in my prospective audience. Abstraction does not come naturally to everyone, but it can be an acquired ability and I hope to make acquiring that ability as smooth as possible. My intention has been to write a book that could be used by a student working alone, though in a discipline such as mathematics solitary study is rarely ideal. There is great advantage in having other interested and motivated students pursuing the same goals with you. You can help each other keep on track and avoid the inevitable pitfalls.

Very few people are able to think in a purely abstract way; we all need various crutches at one time or another, even professional mathematicians. To that end, I use as many diagrams as possible. It is said that a picture is worth a thousand words, and for me at least that is true. The thousand words will eventually get the job done without the picture, but it is a demoralizing experience for all but the most dedicated and talented. It seems foolish to neglect diagrams whenever there may be one available. That said, there is a surprise in store. Pursued far enough, mathematics becomes perhaps the first human intellectual discipline in which one word can be worth a thousand pictures.

This first book is intended to be the first chapter in a series of five on the subject of group theory, but in the end they will all be available separately as well as in a single long volume. A student whose interests lie at only with the first material can escape buying more than they want. This first book introduces as many specific examples of groups as possible organized in families of groups that have particular traits in common. By focusing on their similarities, it is possible to develop many more advanced ideas in a more concrete way. I do not introduce the idea of isomorphisms until the second chapter of the series, which may irritate purists who know that is not the proper way to do this. Proper or not, the focus here is on enough examples with enough variety to display basic algebraic methods and ideas, and to give the student some degree of comfort with the techniques that are required to move on to the next level of abstraction.

Obviously, though, I can't start from ground zero. I have to assume certain information in the reader's background. Mainly I assume that the reader will be familiar with set theory in its most basic form. I assume the reader is familiar with arithmetic, with the various kinds of numbers (real, rational, interger, complex). I assume the reader knows a bit of high school level algebra. And I assume that the reader is familiar with the idea of functions. I also use some elementary facts from combinatorics which the reader must accept on faith as true, or seek out a text on the subject. Otherwise it is definition, theorem, proof, example, and computation just as you might expect from any mathematics textbook. I assume that the reader is familiar with the ideal of mathematical proof and knows the basics of logical thinking, able to distinguish a valid argument from a fallacious one. That said, especially at the beginning of the book, I take a very gentle and gradual approach to proofs to help ease the student into the subject.

But this is a "pure" mathematics book. I do not cover the myriad applications that even something as abstract as group theory invites. In some sense, all mathematics is applied mathematics. But there is mathematics that chooses to ignore the connection of mathematics with reality – *pure* mathematics - and that is what this book does. It has the advantage of focus. It has the disadvantage of focus. That is the way of it.

Mathematics is most like a language, though with a fairly narrow domain of discourse. Mathematics cannot talk about love or justice, but it does an excellent job of talking about what it can. In my opinion, then, mathematics should be pursued like a new language. And like any language, it is essential that the reader learn the definitions and the statements of the theorems by heart. This is more essential in mathematics than in any other language, because mathematical nouns are all abstractions with tenuous connections to the real world. Failure to learn the exact statements of definitions and theorems are the primary cause of failure to understand the subject matter. It is a rookie mistake; don't make it. The definitions as well as the statements of the theorems behave something like the signs Aslan gave to Jill Pole in *The Silver Chair*. Unless you know them perfectly you will not recognize them easily.

Embarking on a study of modern algebra is like embarking on the journey by foot from New York to California. It is like such a journey in two ways. First, algebra is a very large subject. You will never get all the way across its continent, especially if you stop to smell the flowers, as I intend. Second, algebra must be done in the same way as walking to California: one step at a time. Here are the first thirty-one steps; it barely gets us out of Brooklyn. You might decide that you like where you are before you get to California and settle down and that is a perfectly reasonable choice.

I have called this *modern* algebra, which is a common term for this material. What makes it modern, as opposed to "old fashioned" algebra? Algebra, in the West, was really begun by Fibonacci's book in about 1300. What might be called old fashioned algebra developed over five hundred years into a tool with great sophistication, but with no coherent foundation. It was during the 1800's that it was given a foundation as an axiomatic mathematical discipline. The work of such people as Gauss, Abel, Galois, Cayley, Hamilton, Felix Klein, and many others culminated in Burnside's algebra textbook - the first to define and discuss the abstract idea of a group - in the 1890's. In 1800 modern algebra as we have it today did not exist. By 1900, it did exist. During the 20th century it outgrew its infancy and became the mature, and rather overwhelming, creature it now is.

If your only acquaintance with algebra is what you encountered in high school, then I can safely say you have little or no idea what is in store for you here. That is the fun of it. You will learn to think in a new way that you do not yet imagine; you will learn to see new aspects of the world whose existence you had not previously suspected.

Enjoy.

Table of Contents

Lesson 1:	Operations	1
Lesson 2:	Semi-groups	4
Lesson 3:	Monoids	6
Lesson 4:	Inverses	8
Lesson 5:	Groups	11
Lesson 6:	The Fundamental Property of Groups	13
Lesson 7:	Subgroups	15
Lesson 8:	Subgroups Generated by an Element	17
Lesson 9:	Cyclic Groups	19
Lesson 10:	The Lattice of Subgroups	22
Lesson 11:	The Group of Units of a Monoid	24
Lesson 12:	More About U_n	27
Lesson 13:	The Euclidean Algorithm	29
Lesson 14:	Congruence Equations	32
Lesson 15:	Euler's Phi Function	34
Lesson 16:	Primitive Roots, Part 1	37
Lesson 17:	Primitive Roots, Part 2	39
Lesson 18:	The Index	41
Lesson 19:	Permutations and the Symmetric Groups	44
Lesson 20:	Cycles	47
Lesson 21:	Some Cycle Identities	50
Lesson 22:	The Alternating Groups	53
Lesson 23:	Stabilizers	55
Lesson 24:	Dihedral Groups, (odd number of vertices)	57
Lesson 25:	Dihedral Groups, (even number of vertices)	59
Lesson 26:	Dihedral Subgroups of S_n	62
Lesson 27:	Symmetries of the Cube	64
Lesson 28:	Group Presentations	67
Lesson 29:	The Quaternion Group	70
Lesson 30:	The Generalized Quaternions	72
Lesson 31:	The Cayley Digraph	74

Table of Groups	77
Index of Terms	79
Index of Symbols	81

Lesson 1: Operations

If you are reading this book, you have already studied algebra to some extent and have formed a more or less clear idea of what it is about. What you are about to encounter in these lessons will not look very much like what you imagine. Here, we will assemble algebra from its component parts from scratch. We will examine each component in detail and examine how the parts fit together. We will even fit the parts together in non-traditional ways to create new forms of algebra that are unfamiliar.

To say we begin with nothing actually means, mathematically, that we begin with a set. Informally a set is just a list of elements. The *set* is the list itself; the *elements* are the items in the list. Elements can be anything, from actual elephants to abstractions like justice. We will use capital letters to name sets and lower case letters, usually, to name elements. If you know the elements of a set, then you know everything about the set. There is really little more to it. We describe this by saying that a set has no *structure*.

Set theory, as a branch of mathematics, is about relationships between various sets and their subsets and the *functions* which map one set to another. I assume here that you understand the basics of set theory, something about functions, and the basic principles of logic that govern mathematical thinking. You should know, for instance, that the curly braces {} are used to mean that they enclose a list of the elements of a set, either explicitly or by a formula that calculates the elements of the set.

A set is essentially determined by the number of elements it includes. Only rarely do we care about what the elements actually are, whether they are elephants or virtues or letters of the alphabet. The exception is when we have sets of numbers. Mathematics, after all, began with numbers and geometric shapes. But when our interest turns to the elements themselves then our study has gone beyond the realm of set theory and has become something other - in this case, algebra. In algebra we are frequently interested in particular sets of numbers, though algebra also includes sets of other kinds of objects. Some sets of numbers are so important they have special names of their own. Here is a list of many of them.

- $\mathbb{N}$ the set of *natural numbers*, $\{0, 1, 2, \ldots\}$ For historical reasons many mathematicians do not include 0 as an element of $\mathbb{N}$, but it is more convenient to me to include it.
- $\mathbb{Z}$ the set of *integers*, $\{\ldots -2, -1, 0, 1, 2, \ldots\}$, the natural numbers plus the negatives.
- $\mathbb{Q}$ the set of *rational numbers*, $\{p/q \mid p, q \in \mathbb{Z}\}$, that is, the set of fractions.
- $\mathbb{Q}^+$ the set of *positive rational numbers;* note that this does not include 0.
- $\mathbb{Q}^*$ the set of *non-zero rational numbers;* note that this does include negative numbers.
- $\mathbb{R}$ the set of *real numbers*; this is complicated, but essentially it is all decimal numbers even if they do not end after a finite number of digits.
- $\mathbb{R}^+$ the set of *positive real numbers*; no negatives or 0.
- $\mathbb{R}^*$ the set of *non-zero real numbers*; negatives yes, 0 no.
- $\mathbb{C}$ the set of *complex numbers*, $\{a + b\mathbf{i} \mid a, b \in \mathbb{R} \text{ and } \mathbf{i}^2 = -1\}$.
- $\mathbb{C}^*$ the set of *non-zero complex numbers*.
- $\mathbb{I}$ the *closed unit interval*, $\{x \mid 0 \leq x \leq 1\}$.
- $\mathbb{I}^o$ the *open unit interval*, $\{x \mid 0 < x < 1\}$.
- $\mathbb{E}$ the *even integers*, positive and negative.
- $\mathbb{O}$ the *odd integers*, positive and negative.

But algebra also deals with sets of elements that are not numbers. We find our main motivation from sets of numbers, but algebra itself has outgrown its source. The only non-numerical set that we will mention for the moment is:

- $\mathcal{P}(S)$ the *set of all subsets of S*, in mathematical symbols $\{A \mid A \subseteq S\}$

using this particular stylized form of P. The way this is read is "The set of all A such that A is a subset of S." The symbol ⊆ means "is a subset of, or is equal to"; it means that all the elements is A are also in S. The notation tells us first that we are defining a set by the {. Then it gives a variable name for the elements of the set, A. The colon tells us that a formula will follow to tell how to identify those elements, in this case A ⊆ S. Finally } shows us that we are at the end of defining the set. Although algebra takes its inspiration from the numbers, its true objective is to study the kinds of *algebraic structures* that can be imposed on a set. What an algebraic structure is exactly will become more clear as we go.

Algebra begins with arithmetic and we have not yet arrived at arithmetic. We have only spoken of certain sets of numbers, but arithmetic involves procedures like addition, subtraction, multiplication, or division. Procedures like these are the first concepts that we need to introduce, but we must do it abstractly. The procedures that we are looking for must apply equally well to sets of numbers and sets of sets and sets of purely abstract elements whose nature is unspecified. We must define of an arithmetic-like procedure which takes two elements of the chosen set and computes an answer from them, just as addition does in arithmetic. The procedure must also include procedures for combining sets such as unions and intersections of the subsets in $\mathcal{P}(S)$. The following definition fairly abstracts all that is important from the arithmetical procedures, the set procedures, and any other procedure we will study in algebra.

Definition 1: Let S be any non-empty set. An **operation on S** is a function f: S×S → S.

There is a lot in this definition that we must be careful to note. First it tells us that we cannot understand what an operation is unless we remember what a function is. There are three parts to a function: its domain (the input), its range (the output), and the rule which associates elements in the domain with *one* element in the range. This last is an important requirement: each element in the domain must be associated with one and only one element in the range.

So if we encounter a new possible function which we think might be an operation on a set S, how do we proceed to prove it is? Here are four steps

1. The elements of the domain must be ordered pairs of elements of S. In other words, in order to operate, the function must be given a first element and then a second element of S to use in its rule of computation.
2. The computation must be doable for any ordered pair we happen to choose from S.
3. The result of the computation must be some element in S. This mapping back into the set we started from is called *closure* and we say the operation is *closed* on S.
4. The result of the computation must be unambiguous. There can be only one result for a given input.

Once we prove that a possible function satisfies all four of these requirements, then we have proven it is an operation.

Normally we will indicate an operation by putting an operation symbol between the members of the ordered pair rather than writing them like functions. Operation symbols that you already know are +, −, ×, ÷, ∩, and ∪.

Now we already know that addition and multiplication are operations on any of the sets of numbers listed above. Our previous studies of arithmetic have made that clear and we will take these facts for granted without proof. Also our previous studies of set theory will have shown us that union, ∪, and intersection, ∩, are both operations on $\mathcal{P}(S)$. But what about subtraction and division on sets of numbers?

Subtraction is indeed an operation on $\mathbb{Z}$, but subtraction is not an operation on $\mathbb{N}$. The reason is that it is possible to choose an ordered pair of elements of $\mathbb{N}$ and not be able to compute an answer in $\mathbb{N}$. If we choose 2 and then 7 then we compute $2 - 7 = -5$, which is not an element of $\mathbb{N}$. For this same reason, division is not an operation on either $\mathbb{N}$ or $\mathbb{Z}$. It is also not an operation in $\mathbb{Q}$ or $\mathbb{R}$ but for a different reason. It fails on these two because of one element: we are not permitted to divide by 0, so 0 can't be used as the second choice for the ordered pairs we input into the function.

Usually, we will encounter new potential operations when we define them ourselves, and usually we will define such operations by a formula using operations we already know. For example, we may be studying $\mathbb{Z}$ and we may define a function which we hope will be a valid operation on $\mathbb{Z}$. When we define a new operation we will generally use * as the operation symbol. We might define the new operation as $a*b = ab + 1$, where we write the two letters together to symbolize multiplication rather than using the · or the ×. Is * an operation on $\mathbb{Z}$?

First the input for the function is clearly two integers, so that is OK. Second, any two integers are allowed as input since any two integers can be multiplied and then 1 can be added to the result. Third, whatever integers we may substitute for a and b, the answer will certainly also be an integer; we know this because we understand multiplication and addition already. Fourth, we also know that only one answer is possible for each possible ordered pair. Thus, * is a valid operation on $\mathbb{Z}$.

Let's try another possible operation, this time on $\mathbb{Q}$: define
$$a*b = (3a + b)/(a^2 + 2b).$$
The first thing to check is whether $a*b$ is defined for any ordered pair of rational numbers we choose. When an operation is defined using ordinary arithmetic operations on the rationals, the only thing that is problematic is division by 0. Is the denominator ever 0? In this case we see quickly that if $a = 2$ and $b = -2$ then the denominator is indeed 0. Since * is not defined for *all* ordered pairs of rational numbers, it is not a valid operation on $\mathbb{Q}$.

What we have just done with these two examples is check whether or not the potential operations were *well-defined*; this means, did they satisfy the definition? One of the first things to check in any proof is whether everything that is questionable is well defined.

To summarize this lesson, algebra begins when we choose a set, whose elements may or may not be numbers, and then choose an operation for the set. Once we have chosen a set, there are usually many possible operations that can be chosen for it, so the scope of algebra is quite wide. Much too wide. We will have to narrow our focus in order to make much progress.

Exercise 1: Let $S = \{a\}$. List all the possible operations that can be defined on S.

Exercise 2: Let $S = \{a, b\}$. List all possible operations that can be defined on S.

Exercise 3: For each of the fourteen numerical sets named on page 1, check whether or not addition, subtraction, multiplication and division are well-defined operations on them. Make a list of which of these possible 56 combinations of sets and operations are proper studies for algebra.

Lesson 2: Semi-groups

When we choose a set S and one of the possible operations on S to be the algebraic structure we will study, we are left with an embarrassment of riches. There are so many of them, and they are so diverse and many are so complicated in nature that we cannot make headway in analyzing them. We must narrow our focus to something with less variety and complexity.

Since it is choosing the operation that makes the beginning of algebra, it is natural we should narrow our scope by placing some restriction on the operation. Is there a criterion that distinguishes "good" operations from "bad" ones? As it turns out, there is.

Definition 2: Let * be an operation on a non-empty set S. Suppose * satisfies the following property:
$$a*(b*c) = (a*b)*c \quad \forall a,b,c \in S.$$
This property is called the **associative law**, and an operation that obeys the associative law is called an **associative operation**.

The symbol $\in$ means "is an element of". The symbol $\forall$ means "for every". The associative operations are the nice ones in the sense that there is enough order to them that we can make some headway in understanding them. For the time being, we will set aside the non-associative operations. Once we have gotten some substantial insight into the associative operations, our understanding of them will give us the hint we need to begin to understand non-associative operations. Understanding is always serendipitous.

Thus we have a narrower focus: sets with an associative operation defined on them. Now we have arrived at an algebraic structure important enough to have a name.

Definition 3: Let S be a non-empty set with an associative operation * defined on it. Together the set S with the operation *, which we write as $< S, * >$, is called a **semi-group**.

If we have a set S and a candidate * for an associative operation, to prove that $< S, * >$ is indeed a semi-group, we must usually follow a two step procedure:
1. prove that * is a well-defined operation on S.
2. prove that * is associative.

It is the second step that is generally the more difficult one. We must show that the associative law holds for *every* ordered triple of elements in S; if S is very large and we have to check the triples one by one, the task can easily overwhelm us. But when * is defined by a formula it may be much easier to check.

For example, we already know that the operation defined by $a*b = ab + 1$ is a well-defined operation on $\mathbb{Z}$. Is it associative? All we need do is use the formula to check the two sides of the associative law:
$$(a*b)*c = (ab + 1)*c = (ab + 1)c + 1 = abc + c + 1$$
$$a*(b*c) = a*(bc + 1) = a(bc + 1) + 1 = abc + a + 1$$
Hence * is not an associative operation and $< \mathbb{Z}, * >$ is not a semi-group.

We already know some operations from arithmetic that are not associative. For example, on $\mathbb{Z}$ (or any other of the sets of numbers) subtraction is not an associative operation:
$$(7 - 3) - 4 = 4 - 4 = 0 \quad \text{but} \quad 7 - (3 - 4) = 7 - (-1) = 8.$$
Thus we know that $< \mathbb{Z}, - >$ is not a semi-group. Similarly, division is not an associative operation. But we do know from set theory that both union and intersection are associative operations on $\mathcal{P}(S)$. Therefore both $<\mathcal{P}(S), \cap >$ and $<\mathcal{P}(S), \cup >$ are semi-groups.

We will now introduce a new set with an associative operation that will prove to be an invaluable in the rest of our studies. Beginning with the set S as usual, we will now form a new set of all possible functions whose domain and range are both S:

$$F(S) = \{f : S \to S\}$$

The operation we will use on F(S) is the usual composition of functions, which we know already is associative, and for which we will use the usual symbol ∘. We know from previous studies, in calculus if nothing else, that <F(S), ∘ > is a semi-group.

Though we know ∘ is associative, it is instructive to go through the verification. A function is a more complicated entity than a number. In order to show that a function satisfies a certain property, we must show that it does so *for every* element in its domain. Thus, to show that f∘(g∘h) = (f∘g)∘h we must show that [f∘(g∘h)](x) = [(f∘g)∘h](x) ∀x ∈ S. In this case, it goes through very easily because it is defined by a formula. To verify this operation is associative, compute each side of the equation separately and compare.

$$[f\circ(g\circ h)](x) = f((g\circ h)(x)) = f(g(h(x)))$$
$$[(f\circ g)\circ h](x) = (f\circ g)(h(x)) = f(g(h(x)))$$

That is so easy, it looks like a trick. To see more clearly how it works let's take a specific example. Let $f(x) = x^2$, $g(x) = 3x + 2$, and $h(x) = x^3 - 4$. Then

$(g\circ h)(x) = g(h(x)) = g(x^3 - 4) = 3(x^3 - 4) + 2 = 3x^3 - 10$.
and then $[f\circ(g\circ h)](x) = f(3x^3 - 10) = (3x^3 - 10)^2 = \mathbf{9x^6 - 60x^3 + 100}$.
$(f\circ g)(x) = f(g(x)) = (3x + 2)^2 = 9x^2 + 12x + 4$.
and then $[(f\circ g)\circ h](x) = (f\circ g)(h(x)) = (f\circ g)(x^3 - 4) = 9(x^3 - 4)^2 + 12(x^3 - 4) + 4$
$= 9(x^6 - 8x^3 + 16) + 12x^3 - 48 + 4$
$= 9x^6 - 72x^3 + 144 + 12x^3 - 48 + 4$
$= \mathbf{9x^6 - 60x^3 + 100}$.

The example, of course, is not a proof of anything. It is just practice with composition. Never think of a specific case as a proof.

Exercise 4: Of the algebraic structures you found in exercise 3, which of them are semi-groups? Prove your answer.

Exercise 5: Check the following functions to see if they are well-defined operations on the given set; when they are, check to see if they satisfy the associative law on the given set:
  a) $x*y = \sqrt{(xy)}$ on $\mathbb{Z}$. Is there another numerical set this is an operation on?
  b) $x*y = x/(y+1)$ on $\mathbb{R}$. Is there another numerical set this is an operation on?
  c) $x*y = x + y - xy$ on $\mathbb{Z}^+$. Is this an operation on any other numerical set?
  d) $x*y = xy$ on $S = \{n \in \mathbb{Z} \mid n < 0\}$.
  e) $x*y = 3x - 4y$ on $\mathbb{Z}$. Is this an operation on any other numerical set?
  f) $x*y = x + y + xy$ on the set $S = \mathbb{R} - \{-1\}$ (S is $\mathbb{R}$ with the point -1 deleted.)

Exercise 6: On $\mathcal{P}(S)$ define a new operation by this formula: for all A, B ⊆ S,

$$A \# B = (A - B) \cup (B - A)$$

Is $<\mathcal{P}(S), \# >$ a semi-group or not?

Lesson 3: Monoids

Even if we restrict our attention to semi-groups only, there are still too many possible algebraic structures to deal with. Semi-groups have too wide a variety and the scope of the subject is bewildering, particularly when you are beginning from nothing. We need to specialize yet further, to find a sub-class of the semi-groups for the beginning of our study. You might think, having heard of the commutative law in arithmetic, that we will now add the commutative law as well as associativity as a requirement on the operation. That would be reasonable, but it is not what we will do, for reasons that will become clear later.

Instead, we will now distinguish between semi-groups according to whether or not they have a very special kind of element. This is a very different kind of requirement, an existential requirement rather than a universal requirement. It must be handled differently in proofs. Note that since we are still not requiring the algebraic structure to have an operation that is commutative, we must carefully distinguish left from right in any distinction we introduce on operations or on how the elements are operated on.

Definition 4: Let $< S, * >$ be a semi-group. Any element $e \in S$ which satisfies the following equation, **e*s = s** $\forall s \in S$, is called a **left identity** for S. Any element $e \in S$ which satisfies the following equation, **s*e = s** $\forall s \in S$, is called a **right identity** for S. Any element $e \in S$ which is both a left identity and a right identity for S is called a **two-sided identity** for S, or more simply an **identity**.

As soon as we define the existence of a particular kind of thing there are two questions that naturally arise. First, does such a thing even actually exist? Usually we only make such definitions when we have been inspired by actually seeing one. This time the semi-group $< \mathbb{Z}, + >$ provides the example. The element 0 is a two-sided identity for $< \mathbb{Z}, + >$ because $0 + n = n + 0 = n$ $\forall n \in \mathbb{Z}$. Further, the semi-group $< \mathbb{Z}, \cdot >$ also has a two-sided identity, the number 1, because $1 \cdot n = n \cdot 1 = n$ $\forall n \in \mathbb{Z}$. So identities do exist sometimes.

Another question that always occurs is: how many such things might there be? In both $< \mathbb{Z}, + >$ and $< \mathbb{Z}, \cdot >$ there is only one two-sided identity. It is not hard to contrive other examples, though. Consider $< \mathbb{Z}, * >$ where * is defined by $a*b = a$. It is easy to show that * is a well-defined operation on $\mathbb{Z}$. It is also easy to see that any integer n is a right identity for $\mathbb{Z}$ under this operation, so there are infinitely many right identities; but there are no left identities at all. Or consider yet another example, $< \mathbb{Z}, * >$ where * is defined by $a*b = ab^2$. Now we see that there are two right identities, 1 and -1; but there are no left identities.

What about the previously contrived example: $< \mathbb{Z}, * >$ where $a*b = ab + 1$? Under this operation, does $\mathbb{Z}$ have an identity? We can check for the existence of a right identity by asking whether there is any integer x such that $n*x = n$ $\forall n \in \mathbb{Z}$?

If	$n*x = nx + 1 = n$	$\forall n \in \mathbb{Z}$
then	$nx - n = -1$	$\forall n \in \mathbb{Z}$.
We can factor the LHS to get	$n(x - 1) = -1$	$\forall n \in \mathbb{Z}$.
Then	$x - 1 = -1/n$	$\forall n \in \mathbb{Z}$
and	$x = -(1/n) + 1$	$\forall n \in \mathbb{Z}$.

This is clearly impossible, so $< \mathbb{Z}, * >$ has no right identities at all. The same steps show that it has no left identities either.

It seems that the number of identities could be any number. Admittedly, these are contrived examples, but they are valid examples in that they satisfy the definition, and so they must be taken into account. There is no rule in mathematics that allows us to exclude examples just because they seem artificial.

All right. There is considerable variation in what may happen within semi-groups in terms of the number and kind of identities. Is there any restriction or does anything go? The following theorems are easy to prove and narrow the possibilities considerably.

Theorem 1: Let < S, * > be a semi-group and suppose e ∈ S is a left identity and f ∈ S is a right identity. Then e = f.

Proof: The proof is so short if you blink you will miss it. First e = e*f (because f is a right identity). But e*f = f (because e is a left identity). Therefore e = f (because two elements equal to the same element are equal to each other).

This means that a semi-group may have left identities or it may have right identities, but if it has both kinds then they are in fact two-sided identities. Essentially the same proof gives us another restriction on the possibilities.

Theorem 2: Let < S, * > be a semi-group and suppose e ∈ S and f ∈ S are both two-sided identities. Then e = f.

Proof: Follow the same steps as for theorem 1 but with different justifications. First e = e*f (because f is an identity). But e*f = f (because e is an identity). Therefore e = f (because two elements equal to the same element are equal to each other).

This says that when a semi-group has a two-sided identity, it can have only one of them. The two theorems tell us we can classify all semi-groups into four types:

　Type 1: semi-groups with no identities at all – right, left, or two-sided.
　Type 2: semi-groups with one or more left identities, but with no right or two-sided identities.
　Type 3: semi-groups with one or more right identities, but with no left or two-sided identities.
　Type 4: semi-groups with exactly one two-sided identity.

You will not be too surprised that type 4 semi-groups are the ones we plan to focus on. These are the ones that are most like the numerical examples familiar to us from arithmetic. So we have a new definition.

Definition 5: Let < S, * > be a semi-group and suppose e ∈ S is the two-sided identity for S. Then we call < S, * > a **monoid**.

We have already noticed that < $\mathbb{Z}$, + > and < $\mathbb{Z}$, · > are monoids. A note of caution is in order. When a semi-group has an identity and is therefore a monoid, the identity element commutes with every other element in the monoid by definition. *This doesn't mean that the commutative law (which we will define formally shortly) holds in the monoid*, however. No other element of the monoid need commute with anything, but when the identity exists it commutes with everything. In a monoid, there is always a little bit of commutativity.

Exercise 7: Of all the semi-groups you found in exercise 4, determine which ones are monoids and prove your answer. For the monoids, determine the identity.

Exercise 8: Determine if < $\mathcal{P}(S)$, ∩ >, < $\mathcal{P}(S)$, ∪ >, and < $\mathcal{F}(S)$, ∘ > are monoids by finding the identity or proving it does not exist.

Exercise 9: Determine the left, right, or two-sided identities, if any, of the semi-groups found in exercise 5.

Lesson 4: Inverses

A monoid M is a semi-group in which there is a single element, the identity, that is distinguished above every other element in M. The other elements are unremarkable, so far as we know at the moment. But there is a certain inevitability that distinction gets shared around. Once this single special element is found, then there is a natural way to look for patterns or relationships among the other elements that we would never have noticed without the identity element. Once we have the identity element marked out, we naturally should ask: how are the other elements of the monoid related to the identity? Are there differences in how one element is related to the identity than how others are? What kinds of connections can there be between the other elements and the identity?

We must continue to carefully distinguish between left and right uses of the operation. It turns out that the connection between other elements and the identity occurs when we consider pairs of other elements.

Definition 6: Let $< M, * >$ be a monoid and let its identity element be e. If there exist two elements a and b in M such that $a*b = e$ then we say a is a **left inverse** for b and b is a **right inverse** for a. But if $a*b = b*a = e$ so that both a and b are both left and right inverses for each other, then we say that a and b are **two-sided inverses** for each other, or for short, that they are **inverses** of each other.

The existence or non-existence of an inverse of one kind or another occurs element by element. The existence of an identity turned out to be a "global" property in that the identity served for the entire monoid. But an inverse serves only for the element it is paired with. Now notice that we have an unintended consequence of finding the unique identity: it partitions the rest of the set into subsets whose elements are mutually inverses of each other, either left or right, and a single subset of the elements which are not inverses to any other element. Let's see how this partition works in more detail. The next two theorems are parallel to the previous two.

Theorem 3: Let $< M, * >$ be a monoid and e its identity. Let $f \in M$ be any element other than the identity. Suppose $a \in M$ is a right inverse for f and $b \in M$ is a left inverse for f. Then we have $a = b$.

Proof: $a = e*a$ (because e is the identity)
$e*a = (b*f)*a$ (because b is a left inverse for f)
$(b*f)*a = b*(f*a)$ (because * is associative)
$b*(f*a) = b*e$ (because a is a right inverse for f)
and (because e is the identity) $b*e = b$
Therefore $a = b$.

....

This means that a given element of M may have left inverses or right inverses, but if it has both then they equal each other. Now we will use nearly the same proof as theorem 3 to get:.

Theorem 4: Let $< M, * >$ be a monoid and e its identity. Let $f \in M$ be any element other than the identity. Suppose $a, b \in M$ are (two-sided) inverses for f. Then $a = b$.

Proof: $a = e*a$ (because e is the identity)
$e*a = (b*f)*a$ (because b is an inverse for f)
$(b*f)*a = b*(f*a)$ (because * is associative)
$b*(f*a) = b*e$ (because a is an inverse for f)
and (because e is the identity) $b*e = b$
Therefore $a = b$.

....

Thus, an element of a monoid can have only one two-sided inverse. These two theorems allow us to classify *each element* of a monoid, separately and individually, into one of four types. If a is an element of a monoid then either

<u>Type 1</u>: the element a has no inverses at all – right, left, or two-sided.
<u>Type 2</u>: the element a has one or more left inverses, but no right or two-sided inverses.
<u>Type 3</u>: the element a has one or more right inverses, but no left or two-sided inverses.
<u>Type 4</u>: the element a has exactly one two-sided inverse.

A given monoid may have some elements of each type, and this is one reason to think that there may be too great a variety among monoids. We have already mentioned one monoid that illuminates how complicated a monoid may be, namely $< F(S), \circ >$. Let's determine which elements of this monoid are which type.

First, which functions have left inverses? It is helpful to draw a function diagram to help us think. I will not give a formal proof of the following result here, and I will use $\mathbb{N}$ as the underlying set to make the discussion more clear.

Let $f: \mathbb{N} \to \mathbb{N}$ be any function in $< F(\mathbb{N}), \circ >$, and use $i: \mathbb{N} \to \mathbb{N}$ to represent the identity function. For f to have a left inverse means that there must exist a function $g: \mathbb{N} \to \mathbb{N}$ such that $g \circ f = i$. This is an equation that must be satisfied for every element $n \in \mathbb{N}$, that is

$$g(f(n)) = i(n) = n \quad \forall n \in \mathbb{N}.$$

This can only happen if f(n) equals the element that g maps back to n. Since we are trying to determine g we can simply define g to be the function that maps f(n) back to n. In this way we can construct g element by element as f(n) ranges over $\mathbb{N}$. This will work perfectly well as long as f(n) does not map two elements of $\mathbb{N}$ to the same f(n), say f(m) = f(n). Then we are not permitted to map f(n) to both n and m, and so g cannot be defined at that point. This shows that f can have a left inverse *only if* it is injective, that is, one-to-one. When f is injective there seems to be no impediment to defining g at any point.

But what if f is injective but not surjective (that is, onto)? Would that ruin the construction of g? Let's take an example to see how this is not a problem. Let f(n) = 2n. This is certainly an injective function as each natural number has only one double. And f is not surjective since each natural number is mapped to an even number and no natural number is mapped to an odd number. To define g we simply define it to map every even number to half its value. How shall we define g on the odd numbers? Any way at all. However we define g(3), for example, will make no difference to g∘f since f will never map anything to 3. In short, it appears that a function has a left inverse if and only if it is injective, and if it is not surjective then it will have many left inverses.

In order for f to have a right inverse, there must exist a function g such that $f \circ g = i$. Again this must be true for all $n \in \mathbb{N}$:

$$f(g(n)) = i(n) = n \quad \forall n \in \mathbb{N}.$$

Now the identity function i(x) is surjective; every natural number is mapped to itself. Hence if f(x) fails to be surjective we will have problems. Suppose that there is a natural number n such that no natural number is mapped to n by f. Then there is no way to define g(n) so that f will take the result to n; hence g cannot be defined in a way that will make it a right inverse for f. Thus f has a right inverse *only if* it is a surjection.

What if f is surjective but not injective? Let's define the following function: f(n) equals the greatest integer part of n/2. Thus f(3) = 1 and f(2) = 1 so f is not injective. But f is surjective. How then do we define g(1) so that f∘g = i? We can define g(1) = 2 or g(1) = 3 and either choice will work. So if f is a surjection but not an injection it may have more than one right inverse.

It is easy to see then that if f is both injective and surjective – in which case we say f is bijective – then f will have both a left inverse and a right inverse, and, by theorem 3, they are equal to each other and make a two-sided inverse for f.

Thus in general we can expect $< F(S), \circ >$ to have some elements with right inverses (the surjections), some elements with left inverses (the injections), and some with two-sided inverses (the bijections).

Exercise 10: For all the monoids you found in exercises 7 and 9, determine which elements have left inverses, which have right inverses, and which have two-sided inverses.

Exercise 11: Examine $< P(S), \cap >, < P(S), \cup >$ to see which elements have left, right, or two-sided inverses.

Lesson 5: Groups

Monoids are still too diverse for our beginning study. We need to go one more step in narrowing our focus. It is the inverses, naturally, that will enable us to take this last step.

Definition 7: Let < M, * > be a monoid and suppose that every element of M has a two-sided inverse. Then < M, * > is called a **group**.

We have at last arrived at the algebraic structure that will occupy us for the rest of this book. There are many more specialized algebraic structures, many different kinds of groups, and we will consider them in turn. It will be useful, occasionally, to focus on one or the other special sort of group, but we will always return to groups in general.

A caution is in order. For the most part we will not assume that the group operation satisfies the commutative law, but we do know that certain elements always commute with certain other elements. The identity element always commutes with every other element. Each element always commutes with its own inverse. But that is all we can count on.

Yet another caution is in order. Every element has an inverse, but its inverse need not be a different element. It is even fairly common for an element to be its own inverse and in this case a*a = e, where e as usual is the identity element.

The definition of a group has been slow in coming. I have preferred to assemble it slowly rather than give it all at once in a single long definition. When you encounter an unfamiliar potential algebraic structure – basically a set with an operation defined on it, < S, * > – how do you prove it is a group? You must go through a series of steps, just as the definition was given in a series of steps:
  1. Prove < S, * > is a semi-group by proving that * is a well-defined operation on S and is associative.
  2. Prove < S, * > is a monoid by proving that S has an identity with respect to *.
  3. Prove < S, * > is a group by proving that every element of S has an inverse with respect to *.

Many times not all the steps will be necessary because we will have prior knowledge of some of these properties.

We have spoken of algebraic structures and we are beginning to get a glimpse of what this phrase means. The associative operation, with the identity element, and the inverses partitions the underlying set into three kinds of subsets:
  1. the subset {e} consisting in the identity element;
  2. subsets of the form {a, b} where a ≠ b and a and b are mutual inverses;
  3. subsets of the form {c} where c is not the identity and is its own inverse.

All groups are split up like this into a collection of subsets. There is a great deal more to the structure of groups that we will uncover as we proceed.

At this point, as we are feeling our way forward over new and uncertain terrain, we must discuss how we will write about groups. The issue of *notation*, how we write the mathematical concepts we use, is one of those non-mathematical issues that can help or hinder the actual mathematics that we do in critical ways. Mathematical understanding has been held up for centuries sometimes simply because of bad notation. Consider what a great impediment to the mathematical understanding of numbers the Roman numerals were.

There are three main systems we will use to write groups, and at various times we will use all three. You should feel comfortable with all three notations and be able to convert from one to another easily.

First, there is the *additive notation*. In this notation, regardless of the set of elements, we use the usual addition symbol + to represent the operation regardless of what the operation actually is. When we are using additive notation, we will *always* use 0 to mean the identity element and we will *always* use -a to denote the inverse of a.

Second, there is the *multiplicative notation*. In this notation, regardless of the set of elements, we use the usual multiplication symbol · to denote the operation regardless of what the operation actually is. Sometimes we will simply write the elements side by side as we do in high school algebra, so a·b means the same thing as ab. In this notation, we will *always* use 1 to mean the identity element, and we will use a^{-1} or occasionally $1/a$ to denote the inverse of a.

Third, there is the *abstract notation*, which we have frequently used already. In this notation the operation is symbolized by *, the identity element is symbolized by e, and the inverse of a is symbolized by a^{-1}.

There are a few algebraic structures whose notation is so standardized we will continue their use. The composition of functions is always ∘, the union of two sets is always ∪, and the intersection of two sets is always ∩.

We will also continue to use some shorthand that is familiar from arithmetic. When using the additive notation, we use the minus sign to denote the inverse. It is natural to carry this over and use the subtraction sign to denote adding the inverse:

$$a + (-b) = a - b$$

When using multiplicative notation, however, we will usually not use fractions to denote multiplication by an inverse. $a^{-1} \cdot b$ will not be written $(1/a)b$; doing so will cause more confusion than the shorthand will save of space. Only when we can safely assume the commutative property will it prove helpful to use fractions as a way of writing multiplication by the inverse.

The following theorem is one that is easy to prove and which we will use repeatedly.

Theorem 5: In multiplicative notation, $(ab)^{-1} = b^{-1}a^{-1}$.

Proof: The easiest way to prove this is to multiply the RHS by ab and show that the result is the identity.

$$(ab)(b^{-1}a^{-1}) = a(bb^{-1})a^{-1} = a \cdot 1 \cdot a^{-1} = a a^{-1} = 1$$

and $(b^{-1}a^{-1})(ab) = b^{-1}(a^{-1}a)b = b^{-1} \cdot 1 \cdot b = b^{-1}b = 1$

Thus $b^{-1}a^{-1}$ is the inverse of ab.

In words, when we find the inverse of the product of two elements, it is the same as the product of the inverses in the reverse order. The reversing of the order of the elements is due to not having a reliable commutative law.

Exercise 12: State theorem 5 in additive notation and prove it in additive notation.

Exercise 13: Prove that if a group has an even number of elements, then it must have at least one element, not the identity, that is its own inverse.

Exercise 14: Is the set of all fractions in lowest terms with odd denominators a group under +? Is it a group under ·? What if the denominators are even?

Lesson 6: The Fundamental Property of Groups

With the group structure, we have finally arrived at an algebraic structure organized enough that we can make real progress in understanding it. We could specialize even further. There are many different kinds of groups, some relatively easy to understand, and some very complicated, and we will look at many of these kinds of groups as we proceed. But the category of groups is the right category to investigate first. Over the course of the twentieth century, this algebraic structure has been investigated to an incredible depth and we will glimpse only a bit of the understanding that has been achieved.

So far in our study of algebra we have encountered very few equations. Since we have restricted ourselves to structures in which only a single operation is active, equations cannot get as complex as those you dealt with in high school. We will encounter equations like $a*x = b$, but the only ways to make these more complex is to put an exponent on the unknown. Because of these inherent restrictions, equations will not be the main focus of our study here, with one significant exception. Later, once we have introduced a second operation, we will get to the place where equations come into their own, but for now our interest is more basic than that.

However, we know we have arrived at an opportune structure when we discover how easy it is to solve the simple equations we are able to consider. Pause for a moment to think about what tools we have for solving even this simple equation. Not many. We can use the associative law, we can use the existence of inverses and of the identity, and we can "star" both sides of the equation by the same element. That is it. But these are enough to make a good start.

Theorem 6: (The Cancellation Laws) Let $< G, * >$ be a group. Then the equation $a*x = b$ has the unique solution $x = a^{-1}*b$, and the equation $x*a = b$ has the unique solution $x = b*a^{-1}$

Proof: We have to do two cancellation laws, one from the right and one from the left, because groups may not be commutative. We will only do the proof for the left cancellation law here because the steps will be the same for the right cancellation law.

First the existence of the solution. Multiply both sides of $a*x = b$ by a^{-1} from the left. We will get $a^{-1}*(a*x) = a^{-1}*b$. Note that we have to multiply both sides from the *left* by a^{-1}. It would not be valid to multiply one side from the left and one side from the right unless we knew * satisfied the commutative law. Now use the associative law on the LHS to get the equation $(a^{-1}*a)*x = a^{-1}*b$. By definition of the inverse, the LHS becomes $e*x = a^{-1}*b$. Then by definition of the identity, we have $x = a^{-1}*b$, and we know a solution exists because we have found it.

Now we need to prove uniqueness of the solution. Uniqueness is usually proven by contradiction, and that's what we use here. Suppose $x = c$ and $x = d$ are both solutions of $a*x = b$. Then $a*c = b$ and $a*d = b$ are both true. But then $a*c = a*d$. Now multiply both sides of this equation *from the left* by a^{-1}: $a^{-1}*(a*c) = a^{-1}*(a*d)$. Now we can use the associative law to get $(a^{-1}*a)*c = (a^{-1}*a)*d$. Then $e*c = e*d$ and by definition of the identity $c = d$. Thus our original supposition requires that there be only one solution.

The cancellation laws arise directly from the guarantee that every element has an inverse. We had to repeatedly narrow our focus until we reached the point at which inverses existed in order to get to the cancellation laws. But inverses can't exist unless there is an identity, so we had to get that one first, so we had to get to monoids before we got to groups. And everything hinged, from the very beginning, on being able to count on the associative law. Nearly every proof we give will use the associative law at one point or another. That is why we began with semi-groups. It makes sense in retrospect. We needed to assimilate all the properties that would result in the cancellation laws being valid, and groups are the point at which they became valid for the first time.

Since we are now able to talk about equations, there is a notational convention that can be quite confusing at first. We will adopt a shorthand way of writing things that should be approached with caution. It will frequently be necessary at various times to have an element repeatedly operate on itself, just as we do in arithmetic, and we adopt the same shorthand here as in arithmetic. It looks different in the different styles of notation, however.

Consider additive notation first. Let $<G, +>$ be a group. We will often form repeated sums of an element with itself: g + g + g + g + g + g + g + g + g + g + g + g + g + g + g + g + g + g etc. This can go on for quite a stretch and we need a briefer way of writing it just as we needed in arithmetic, and we will use the same convention. The above sum will be written as 18g. We write it this way even though 18 is a natural number and not usually an element of the group G. We trust you to know, from your previous understanding of arithmetic, what something like 18g must mean. When you see ng, and you know numerical sets are not involved with the group under consideration, you should automatically interpret it according to this convention.

In multiplicative notation, the shorthand involves the use of exponents. Let $<G, \cdot>$ be a group. Then we will write g·g·g·g·g·g·g·g·g·g·g·g·g·g·g·g·g·g·g·g·g·g·g·g more conveniently as g^{24} and remember what it means from arithmetic.

Having introduced these conventions, the problem for the unwary student is bringing the usual tools of arithmetic along with the conventions. I mean the distributive law in additive notation, and the laws of exponents in the multiplicative notation. Again, because we cannot count on the commutative law, these extra tools are only partially valid.

Theorem 7: Let $<G, +>$ be a group and $g, h \in G$. Then $ng + mg = (n + m)g$.

Proof: The LHS means that n copies of g are all added together and then another m copies of g are added to them. The result is that n + m copies of g are added together, which is what the RHS says.

However, the other half of the distributive law does not work: $mg + mh \neq m(g + h)$. You will be asked to give reasons why this doesn't hold in the exercises.

Exercise 15: Restate theorem 7 in multiplicative notation and prove it in that notation.

Exercise 16: Show why $mg + mh \neq m(g + h)$. Restate this in multiplicative notation.

Exercise 17: Restate theorem 6 in both additive and multiplicative notation.

Exercise 18: Let S be any finite non-empty set. Prove that the set of bijections $f: S \to S$ is a group.

Lesson 7: Subgroups

So far we only know a small amount about the structure of a group. We know that a group $< G, * >$ is partitioned into subsets: the subset $\{e\}$, the subsets $\{a, a^{-1}\}$, and the subsets $\{b\}$ where b is its own inverse. But there is a lot more to the structure of a group and now we will begin to uncover some of it. In this lesson we will examine some subsets of the group that structurally are of essential importance. These are the subsets that are self-contained in the sense that they are themselves groups under the same operation as G. These subsets are self-contained in the sense that if we threw away everything else in G, the subset that was left would still be a group.

Definition 8: Let $< G, * >$ be a group, and let $H \subseteq G$. If $< H, * >$ is itself a group then H is called a **subgroup** of G, written $H \leq G$. In the uninteresting case when $H = G$, we say H is the **improper subgroup**. At the other extreme, when $H = \{e\}$, we say that H is the **trivial subgroup**. All other subgroups besides the improper subgroup and the trivial subgroup are called **proper subgroups**.

As you see, we typically suppress any reference to the operation itself and just refer to the group G or the subgroup H, though technically these are only sets unless we specify the operation. We will specify the operation only when it is necessary for clarity. We sometimes abuse notation even more by referring to the trivial subgroup as e (or 1 or 0, depending on the notation we are using), as if it were the element itself that was the subgroup rather than the subset composed of the element. It is a subtle difference, but it is a real difference.

How do we prove that a subset is a subgroup? We need not worry about whether the operation on the subset is associative since it will inherit that property from the group. Clearly any subset that does not include the identity can't be a subgroup. Clearly if g is in the subset, then g^{-1} must also be in the subset. And clearly, if both g and h are in the subset, then both gh and hg must also be in the subset. We can check each of these requirements to prove the subset is a subgroup, but there is a shorter path that will get us there. We will now switch to multiplicative notation.

Theorem 8: Let $< G, \cdot >$ be a group and let $H \subseteq G$. Then $H \leq G$ iff whenever $g, h \in H$ then $gh^{-1} \in H$.

Proof: $\Rightarrow$) Suppose $g, h \in H$ and $H \leq G$. Since H is itself a group and $h \in H$, then by definition $h^{-1} \in H$. Since H is a group and both g and $h^{-1} \in H$, by definition of a group, $gh^{-1} \in H$.

$\Leftarrow$) Suppose that for all $g, h \in H$ we know $gh^{-1} \in H$. Since this is true *for all* g and h in H, we may choose $g = h$. Then $hh^{-1} = 1 \in H$. Now we know $1 \in H$, we may choose $g = 1$, so $1 \cdot h^{-1} = h^{-1} \in H$. Thus H includes the inverses of all its elements. Since $h^{-1} \in H$, we now know $g(h^{-1})^{-1} = gh \in H$, so H includes the products of any two its elements. The associative law holds in H since all the elements of H are also elements of G and the associative law holds in G. Therefore H is itself a group under the same operation as G.

To prove a subset H of G is a subgroup, choose two arbitrary elements of H and show that the subset also includes the first one times the inverse of the second. It's a one step procedure.

Not all groups have proper subgroups, but it is rare when one does not. Subgroups do have one special property that we will use repeatedly in our future investigations.

Theorem 9: Let $< G, * >$ be a group and suppose both $H, K \leq G$. Then $H \cap K \leq G$.

Proof: According to theorem 8, all we need to show is that if $g, h \in H \cap K$ then $gh^{-1} \in H \cap K$. Now if we take $g, h \in H \cap K$, then $g, h \in H$ and $g, h \in K$, by definition of the intersection. Further $g, h \in H$ implies $gh^{-1} \in H$ by theorem 8, and similarly $g, h \in K$ implies $gh^{-1} \in K$. Since $gh^{-1} \in H$ and $gh^{-1} \in K$, then $gh^{-1} \in H \cap K$ by definition of the intersection. Therefore by theorem 8 we have $H \cap K \leq G$.

The intersection of two subgroups is a subgroup. However, we do not fare so well with the union of two subgroups. Consider $H \cup K$. In general the union includes elements, say h, that are in H but not in K and some, say k, that are in K but not in H. For the product hk, there is no guarantee that it will be in either H or K. In short, the union of two subgroups may not be a subgroup. We will see a specific example of this before long.

However there are certainly subgroups of G that contain H ∪ K; for example, the improper subgroup G does. Collect together all the subgroups of G that do contain H ∪ K. If we take the intersection of these subgroups, (and we are talking about finite groups here so there can only be a finite number of subgroups) we will still have a subgroup that contains H ∪ K, and it will be the smallest subgroup that does contain H ∪ K as you will prove in the exercises. I am brushing some technicalities under the rug here. We have not shown more than that the intersection of *two* subgroups is a subgroup, but it follows easily that all finite intersections of subgroups are subgroups. This is more problematic for an intersection of an infinite number of subgroups, but we will not have to deal with infinite intersections and so we will set aside the question. Meanwhile, we have the following new terminology.

Definition 9: Let $<G, *>$ be a group and $H, K \leq G$. Then the **join** of H and K, which we will write as $H \vee K$, is the intersection of all the subgroups of G which contain both H and K.

One more definition is appropriate here. It is always worth knowing the total number of elements in a group, so:

Definition 10: Let $<G, *>$ be a group. The number of elements in the set G is called the **order** of G. If G is an infinite group, its order is equal to the cardinality of the set G. We denote the order of G by absolute value signs, $|G|$.

As we go we will keep a list of the groups we know arranged according to their order. This will give us more insight into the great variety that exists within the group structure.

Exercise 19: Restate theorems 8 and 9 and their proofs in additive notation.

Exercise 20: Let F be the group of bijections $f: S \to S$. For a fixed element $s \in S$, let E be the subset of all the bijections of F which fix s; that is, for which $f(s) = s$. Show that $E \leq F$.

Exercise 21: Show that the join of two subgroups of G is the smallest subgroup of G that contains them both.

Exercise 22: Show that the set of all square roots of elements of $\mathbb{Q}$ is not a subgroup of $\mathbb{R}$.

Exercise 23: Show that $\{q \in \mathbb{Q} \mid nq = 2 \text{ for some } n \in \mathbb{Z}\}$ is not a subgroup of $\mathbb{Q}$.

Lesson 8: Subgroups Generated by an Element

When we are trying to understand a given group, it is of first importance to know what its subgroups are. But finding them all can be quite a difficult job. Fortunately the most basic subgroups are very easy to find and understand. It all starts with choosing an individual element. Let $<G, \cdot>$ be a group and let $g \in G$ be any element other than 1. Since we are using multiplicative notation, we will collect together all the powers of g into a single set and denote it $<g>$:

$$<g> = \{g^n \mid 1 \leq n\} = \{g, g^2, g^3, \ldots\}$$

In additive notation we would have collected together all the multiples of g into a single set and it would have been written like this:

$$<g> = \{ng \mid 1 \leq n\} = \{g, 2g, 3g, \ldots\}$$

It is relatively easy to prove the next theorem, for finite groups. We will postpone considering infinite groups, so unless otherwise stated all groups, except the numerical ones we have already considered, will be finite.

Theorem 10: Let $<G, \cdot>$ be a (finite) group and let $g \in G$ be any element other than the identity. Then $<g> \leq G$.

Proof: To prove a set is a subgroup it is quickest to use theorem 8 but proving this the long way will help us understand better what $<g>$ is.

First, is multiplication well defined on the set $<g>$? Since the only elements of $<g>$ are powers of g, this is asking, "Is the product of any two powers of g another power of g?" We have already noted that the answer is yes and that the law of exponents applies. We automatically know that multiplication is associative since it is the group operation.

Second, is the identity in $<g>$? This is easy to see if $|G| < \infty$. As we compute succeeding powers of g, at some point there must be a power of g that is equal to a previous power, because there are only finitely many elements available. Suppose $g^k = g^m$ and we may as well assume that $k > m$. Then multiply by the inverse of g^m on both sides of the equation. That inverse is given by g^{-m}; it is certainly an element of G though we do not know yet if it is in $<g>$ because we included only positive powers of g in $<g>$ in the definition. In any case, when we multiply both sides by g^{-m} we get $g^{k-m} = 1$. Since $k - m > 0$ we have a positive power of g, an element of $<g>$ that equals the identity.

Finally, does $<g>$ contain the inverses of each of its elements? Choose any element of $<g>$, say g^m, and suppose k is the power of g that equals 1, which we now know must be in the set $<g>$. There are two cases:

Case 1: If $m < k$ then $k - m > 0$ and $g^{k-m} \in <g>$. Now $g^m g^{k-m} = g^k = 1$. Thus $<g>$ includes the inverse of g^m and that inverse is g^{k-m}.

Case 2: if $m > k$, then $m - k > 0$ and therefore $g^{m-k} \in <g>$. So $g^m = g^{m-k}g^k = g^{m-k} \cdot 1 = g^{m-k}$. Similarly $g^{m-2k} = g^m$. Repeat this process until we at last reach the first multiple of k for which $0 < m - qk < k$ and we know $g^{m-qk} = g^m$. By the previous case, the inverse of g^{m-qk} is an element of $<g>$, and its inverse is given by $g^{k-(m-qk)} = g^{(q+1)k-m}$.

We need not consider the case $m = k$ since that is the identity. Therefore $<g>$ is a subgroup of G.

It is important to notice is that if $g^k = 1$, then $g^{m+nk} = g^m$ for any n. It should be obvious in retrospect. Theorem 10 justifies introducing some new terminology.

Definition 11: Let $<G, \cdot>$ be a group and let $g \in G$. $<g>$ is called **the subgroup generated by g** and the element g is said to be a **generator** for $<g>$.

This naturally leads to more terminology. We have already defined the order of a group G as the number of elements in the group. Now we define the order of an element g in the group.

Definition 12: Let $<G, \cdot>$ be a group and let $g \in G$. The smallest positive integer n for which $g^n = 1$ is said to be **the order of the element g**. If no such integer exists, we say that g has **infinite order**.

We saw in the proof of theorem 10 that once a power of g equals the identity, that higher powers of g just recycle through the previously computed values. Thus, if $g^n = 1$, then the entire subgroup generated by g consists in the elements $\{g, g^2, g^3, \ldots, g^{n-1}, 1\}$. In other words, the subgroup generated by an element of order n has order n. Definition 12 is a natural extension of definition 10.

Let's consider some examples. Consider $<\mathbb{Z}, +>$. In additive notation, collecting the powers of an element becomes collecting the multiples of an element. Thus we can easily compute:

$<2> = \{\ldots, -4, -2, 0, 2, 4, 6, 8, \ldots\}$ This subgroup is an infinite subgroup since no multiple of 2 is ever equal to 0. Since it is an infinite group we must include the negative multiples of 2, which are all the inverses, as well. This means our definition of $<g>$ must be modified for infinite groups.

$<5> = \{\ldots, -10, -5, 0, 5, 10, 15, 20, \ldots\}$ again an infinite subgroup. What is $<2> \cap <5>$? A little thought should show you that $<2> \cap <5> = <10>$ since $<10>$ includes all the integers that are both multiples of 2 and multiples of 5. What is $<2> \vee <5>$? The answer is part of an exercise.

Now let's consider the group $<\mathbb{Q}^+, \cdot>$, again an infinite group. This time it is multiplicative notation so we collect all the powers of the element *including the negative powers* which we must have to get the inverses:

$<2> = \{\ldots, \frac{1}{8}, \frac{1}{4}, \frac{1}{2}, 1, 2, 4, 8, 16, \ldots\}$

$<5> = \{\ldots, 1/25, 1/5, 1, 5, 25, 125, \ldots\}$ This time $<2> \cap <5> = 1$, the trivial subgroup. What is $<2> \vee <5>$?

One more example but with a finite group. . It was an exercise to show that the set of all bijections of a set onto itself forms a group. Take $S = \{a, b, c, d\}$ and consider the bijection f defined by the first function diagram. The operation is composition, so to find $<f>$ we must repeatedly compose f with itself until we reach the identity function. It is not difficult to compute $f \circ f = f^2$ (given in red). Then we must take $f^2 \circ f = f^3$ (again shown in red). Since $f^3 = i$ the function f is order 3 in the group of bijections of S and $<f> = \{i, f, f^2\}$.

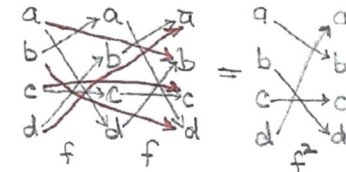

Exercise 24: Prove that if $K \leq H$ and $H \leq G$ then $K \leq G$.

Exercise 25: Find two specific subgroups of $<\mathbb{Z}, +>$ whose union is not a subgroup.

Exercise 26: Prove that if g has order n, then $g^{-1} = g^{n-1}$.

Exercise 27: Show that in $<\mathbb{Z}, +>$, for n and $m \in \mathbb{Z}$ we have $<n> \cap <m> = <lcm(n,m)>$ (where lcm stands for least common multiple) and $<n> \vee <m> = <gcd(n,m)>$ (where gcd stands for greatest common divisor).

Exercise 28: Let G be any group and $x \in G$. Then x and x^{-1} have the same order.

Exercise 29: Let G be any group and $x, y \in G$. Show that xy and yx have the same order.

Exercise 30: Let G be any group and $x \in G$. If x has order $n = st$, show that x^s has order t.

Lesson 9: Cyclic Groups

We have now arrived at the simplest type of algebraic structure of all, the cyclic groups. Note that I will now quit specifying the operation on the group unless there is a reason to.

Definition 13: Let G be a group. If there exists an element g ∈ G such that < g > = G, then G is called a **cyclic group**.

A cyclic group consists in a single element and all its multiples or, in multiplicative notation, all its powers. We will use additive notation now. The order of the cyclic group is the same as the order of its generator. We are already familiar with one cyclic group. Until you are accustomed to the notation we will denote elements of cyclic groups with boldface. Use **1** to name the generator of the group and let it have order 12. In other words, twelve **1**'s, or **1**+**1**+**1**+**1**+**1**+**1**+**1**+**1**+**1**+**1**+**1**+**1** = **12**, equals the identity of the group. In additive notation is the identity is **0**, so **12** = **0**. Hence the elements of the group are {**0, 1, 2, 3, 4, 5, 6, 7, 8, 9, 10, 11**}. The addition of two elements proceeds as usual except that every time we get to **12** we can change it to **0**.

This algebraic system is one you are already familiar with, possibly under the name "clock arithmetic". On our clocks we use 12 o'clock instead of 0 o'clock, but otherwise it is the same. The equation **11** + **1** = **0** means that one hour after eleven o'clock is twelve o'clock, which we will call **0**. In the same way two hours later than eleven o'clock is one o'clock, or **11** + **2** = **1**. The arithmetic in this group can be understood as ordinary addition except that instead of occurring on a number line it occurs on a number circle. Adding is represented schematically as taking steps clockwise, and adding an inverse – subtracting – is represented schematically as taking steps counter-clockwise. Clock arithmetic is the perfect picture of the working of a cyclic group of order 12 in additive notation.

But it is perfectly clear that there is nothing magical about the number 12. We could have done "clock arithmetic" with any number of hours going around the clock. Thus, changing the size of the clock gives us a picture of a cyclic group of any order we choose. Thus we have a whole family of groups, the cyclic groups, with one cyclic group of each order from 2 on up. A cyclic group of order 1 would just be the trivial group. We will use a special symbol to denote the cyclic group of order n, namely Z_n. The ordinary clock arithmetic you are accustomed to is thus an example of Z_{12}.

It is common to use numerals to signify the elements in a cyclic group when additive notation is used, so let me repeat that for a few lessons numerals in boldface will indicate elements in some cyclic group, and numerals not in boldface will indicate ordinary integers. So we will write 7 + 8 = 15 always, but **7** + **8** = **3** in Z_{12} or **7** + **8** = **5** in Z_{10}. This notation will be a big help in avoiding confusion for the first few lessons about cyclic groups, but we will drop it later.

It is seldom that a group is produced by a single generator. Cyclic groups are rare, but important. In a sense, they turn out to be the "atoms" out of which groups are made, analogous to primes being the "atoms" of all natural numbers. At this point we are ready to introduce formally what we have mentioned from time to time already.

Definition 14: Let * be an operation defined on some set S. If a*b = b*a ∀a, b ∈ S then * is called a **commutative operation**. The rule a*b = b*a ∀a, b ∈ S is called the **commutative law**.

Definition 15: Let G be a group whose operation is commutative. G is called an **Abelian group**.

We can now prove a series of easy theorems detailing the essential structure of cyclic groups. We will do these in additive notation and I will have you translate them into multiplicative notation in the exercises.

Theorem 11: Let $<G, +>$ be a cyclic group. Then G is an Abelian group.

Proof: If G is a cyclic group, there is some element **g** ∈ G such that $G = <\mathbf{g}>$. If we pick two arbitrary elements of G, say **a** and **b**, each of them must equal some multiple of **g**. Then we can let $\mathbf{a} = n\mathbf{g}$ and $\mathbf{b} = m\mathbf{g}$. So

$$\begin{aligned}
\mathbf{a} + \mathbf{b} &= n\mathbf{g} + m\mathbf{g} \\
&= (n + m)\mathbf{g} && \text{(by theorem 7)} \\
&= (m + n)\mathbf{g} && \text{(because addition of integers is commutative)} \\
&= m\mathbf{g} + n\mathbf{g} && \text{(by theorem 7)} \\
&= \mathbf{b} + \mathbf{a}.
\end{aligned}$$

Theorem 12: Let $<G, +>$ be a cyclic group. Then every subgroup of G is cyclic.

Proof: We will prove this by contradiction. Let $<G, +>$ be cyclic, $H \leq G$, and *suppose H is not cyclic*. Let **g** be the generator of G. Since H includes only positive multiples of **g**, there is a smallest positive multiple of **g** that is in H; called it n**g**. Since H is assumed to be not cyclic, there must be some multiple of **g** that is not in $<n\mathbf{g}>$ but is in H; call this m**g**.

Since m**g** is not a multiple of n**g**, m must be between two consecutive multiples of n, so let's say $kn < m < (k+1)n$. Now since n**g** ∈ H, kn**g** ∈ H and so is its inverse, –kn**g** ∈ H. Since H is a subgroup, the element $m\mathbf{g} + (-kn)\mathbf{g} = (m - kn)\mathbf{g}$ is also in H. However $0 < m - kn < n$. This is a multiple of **g** in H smaller than the smallest multiple of **g** that is in H, a contradiction.

Therefore every subgroup of G is cyclic.

Theorem 13: Let $<G, +>$ be a cyclic group generated by the element **g** which is of order n; and let $1 < m < n$, $H = <m\mathbf{g}>$, and let $k = \text{g.c.d.}(n,m)$. Then $|H| = n/k$.

Proof: Since $k = \text{g.c.d.}(n,m)$, we know $n = kn'$ and $m = km'$ for some n' and m'. Then the order of m**g** is at most n/k because

$$(n/k)m\mathbf{g} = (nm/k)\mathbf{g} = ((n'km'k)/k)\mathbf{g} = n'km'\mathbf{g} = nm'\mathbf{g} = m'(n\mathbf{g}) = m'\mathbf{0} = \mathbf{0}.$$

It is left to the exercises to prove the order of m**g** is at least n/k. Therefore the order of m**g** equals n/k.

The following corollaries are immediate consequences of what we now know.

Corollary 13a: Let $<G, +> = <\mathbf{g}>$. Let **g** be of order n, and let m be relatively prime to n. Then $<m\mathbf{g}> = G$.

Corollary 13b: Let G be a cyclic group of order n. If m|n then G has exactly one subgroup of order m.

Corollary 13c: Let G be a cyclic group of prime order. Then G has no proper subgroups.

As it turns out, *the groups $\mathbf{Z}_p$ where p is a prime are the only groups that have no proper subgroups*. Every other group is more complicated, but thanks to corollary 13b we now know exactly what subgroups $\mathbf{Z}_m$ has and how many copies there are of each, namely one. Thus we have already arrived, after very little work, at – theoretically - a complete understanding of all cyclic groups. A complete understanding, yes, but there are still useful things to prove. One more theorem, then, to end the lesson.

Theorem 14: Let $<G, +> = <\mathbf{g}>$, and let $H = <h\mathbf{g}>$, $K = <k\mathbf{g}>$ be two subgroups. Then

$$<h\mathbf{g}> \cap <k\mathbf{g}> = <n\mathbf{g}> \text{ where } n = \text{l.c.m.}(h,k) \text{ and}$$
$$<h\mathbf{g}> \vee <k\mathbf{g}> = <m\mathbf{g}> \text{ where } m = \text{g.c.d.}(h,k).$$

Proof: We know that $<h\mathbf{g}> \cap <k\mathbf{g}>$ is a subgroup and so by theorem 12 it is cyclic and generated by some multiple of g, say n**g**. Since n**g** ∈ $<h\mathbf{g}>$ and n**g** ∈ $<k\mathbf{g}>$, we know n is a multiple of h and a multiple of k, in other words, n is a common multiple of h. But any common multiple of h and k will clearly be a multiple of g that is in both $<h\mathbf{g}>$ and $<k\mathbf{g}>$ and thus in their intersection. Therefore the multiples of n must include all the common multiples of h and k, and so n must be the least of them.

Now for the second statement. We know that $\langle h\mathbf{g} \rangle \vee \langle k\mathbf{g} \rangle$ is the smallest subgroup that contains both $\langle h\mathbf{g} \rangle$ and $\langle k\mathbf{g} \rangle$. Call its generator $m\mathbf{g}$. $m\mathbf{g}$ must generate both $h\mathbf{g}$ and $k\mathbf{g}$, that is, h must be a multiple of m and k must be a multiple of m. Said in the opposite way, m must be a divisor of both h and k; that is, m must be a common divisor of h and k. But the join of two subgroups is the smallest subgroup containing them both, and so m must be the largest possible such common divisor in order to generate the fewest number of elements. Therefore we can conclude m = g.c.d.(h,k).

We now know all the subgroups of $\mathbf{Z}_m$, that there is only one subgroup of each order, and we know which element of $\mathbf{Z}_m$ generates it. We now need to devise a way to put all of this knowledge together into an easily usable form. That will be the subject of the next lesson.

Exercise 31: Finish the proof of theorem 13.

Exercise 32: List all the possible generators of $\mathbf{Z}_{100}$. Find $\langle 24 \rangle \cap \langle 52 \rangle$ and $\langle 24 \rangle \vee \langle 52 \rangle$.

Exercise 33: Give proofs of each of the corollaries of theorem 13.

Exercise 34: Let G be a group, and let $x \in G$. Show that if x has order n, then 0, x, 2x, 3x, …, (n–1)x are all distinct.

Abelian groups are named after Niels Henrik Abel, who was born in 1802 in Nedstrand, Norway. Abel showed an early aptitude for mathematics and by the time he was at the university in Oslo he had already surpassed all the mathematical education available in Norway. To continue his education, he went to Copenhagen where he did much of his most fundamental research. Norway was something of a backwater at the time and Abel was on the fringe of the mathematical community. As a result his truly important contributions to mathematics were neglected during his life. In 1823 he was the first to give a complete proof of the insolvability of the general fifth degree polynomial equation (a topic we will discuss in a future chapter). When it was published in 1826 no one noticed it. While traveling in Europe hoping to meet other mathematicians and gain a foothold in the mathematical community, he contracted tuberculosis in 1826 and died by the age of 27. After his death the fundamental importance of his contributions were finally recognized.

Lesson 10: The Lattice of Subgroups

We will continue to use the additive notation for cyclic groups and we will continue to use boldface numerals to represent elements of the group. Thus we are now thinking of Z_m as the cyclic group with **0** as the identity and with **1** as the generator. This means we are thinking of Z_m as being the set {**0, 1, 2, 3, ..., m – 1**} under the operation + and remembering the rule that whenever we get to **m** we can replace it with **0**. There is one particular advantage we can gain by using additive notation that we will exploit in the next lesson, but in general cyclic groups are frequently written with multiplicative notation.

What we proved in lesson 9 can be boiled down to two facts: first, every divisor **k** of **m** generates a subgroup of order m/k; and second, since this is true of *every* divisor, Z_m has exactly one cyclic subgroup of order k for every divisor **k** of **m**. Knowing all this permits us to list every subgroup of Z_m quite easily unless **m** is a very large number. It also enables us to draw an especially helpful diagram of all the subgroups of Z_m showing their relationships to each other. This is called a *lattice diagram* and it is invaluable in giving insight into the internal working of a group.

The rules for the construction of a lattice diagram are the following:

1. Arrange the subgroups of Z_m vertically according to their size. Z_m will be at the top and **0** will be at the bottom. Spread them out enough vertically that it is unambiguous which is larger and which is smaller.

2. Draw a line, straight if possible, from each subgroup to the next larger subgroups. The idea is that there should be only one path from one subgroup up to a larger one containing it. If there is an intermediate subgroup, go to it with one line and above it with another line; do not put an additional line from the smaller subgroup to the larger because that would give a second path and clutter up the diagram unnecessarily. For example, if Z_2, Z_4, and Z_8 are all subgroups of Z_m, there should be a line from Z_2 to Z_4, and a line from Z_4 to Z_8 but no line from Z_2 to Z_8. Z_2 is a subgroup of Z_8 *through* being a subgroup of Z_4.

3. Spread the subgroups out horizontally so that the lines you have drawn are as untangled as possible.

Drawing a lattice diagram is something of an art form. It is worth the effort to make several attempts at drawing the neatest diagram possible because a sloppy diagram can be as hard to understand as having no diagram at all. There are groups that have so many subgroups with so many internal relationships that drawing a readable lattice diagram is nearly hopeless. Some of the groups of order 32, for example, are nearly impossible. Still, in many cases a lattice diagram provides a helpful visual representation of the internal structure of the group. Even when the group is too complex for a lattice diagram, a partial lattice diagram is usually helpful.

The first step in constructing a lattice diagram is to make a list of all the divisors of the order of the group you are diagramming. This can be a bit difficult if the order of the group is very large, but with a bit of care it can be done even for fairly large orders. It helps to know the prime factorization of the order. If no more than three prime factors are involved in the order of the group, a lot of clarity can be obtained by assigning a direction to each prime. In the diagram of Z_{30} given below, the lines that go straight up indicate multiplying the order by 5, the lines that go up and to the right indicate multiplying the order by 3, and the lines that go up and to the left indicate multiplying the order by 2. In this way we can easily work our way up from **0** to Z_{30} and not leave out any subgroup. If more than three primes are involved then a new device must be invented to help the diagram be coherent.

Notice that the lattice diagram shows quickly what the intersections and joins of two subgroups are. To find the intersection of two subgroups, follow the lines down from each of the two subgroups until they first connect. Thus the intersection of Z_{15} and Z_{10} is easily seen to be Z_5, which we knew already since 5 is the greatest common divisor of 15 and 10. The intersection of Z_{15} and Z_2 is the trivial subgroup 0. To find the join of two subgroups, follow the lines upward from the two groups until they first connect. We will conclude this lesson with a theorem which is interesting in its own right though we will not often make use of it.

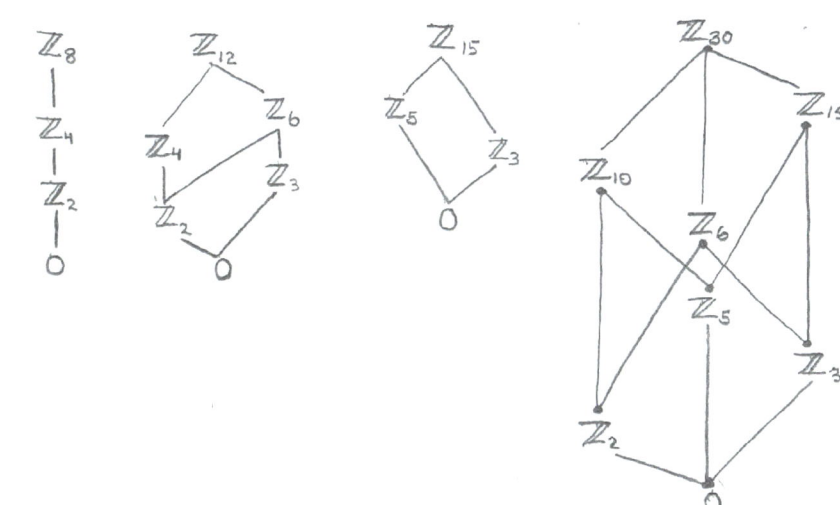

Theorem 15 (The Modular Law): Let G be a cyclic group and let J, H, and K be any subgroups of G. Then we have J ∩ (H ∨ K) = (J ∩ H) ∨ (J ∩ K).

Proof: We will use theorem 13. Suppose J = < **j** >, K = < **k** > and H = < **h** >. Let **d** be the g.c.d. of all three **j**, **k**, and **h**. Next let **pd** = g.c.d. (**j,k**), **qd** = g.c.d. (**h,k**), and **rd** = g.c.d. (**j,h**). Then we know that **p**, **q**, and **r** are relatively prime, and that **j** = **apdr**, **k** = **bpdq**, and **h** = **cqdr**, where **a, b,** and **c** are mutually relatively prime. Now we will calculate the generator of the two sides of the modular law separately.

First the LHS: the generator of H ∨ K is the greatest common divisor of **cqdr** and **bpdq**, which is **dq**. Then the generator of J ∩ (H ∨ K) is the least common multiple of **dq** and **apdr**, which is **apdqr**. Hence J ∩ (H ∨ K) = < **apdqr** >.

Now the RHS: the generator of J ∩ H is the least common multiple of **apdr** and **cqdr**, which is **apcqdr**; the generator of J ∩ K is the least common multiple of **apdr** and **bpdq**, which is **abpqdr**. Finally the generator of (J ∩ H) ∨ (J ∩ K) is the greatest common divisor of **apcqdr** and **abpqdr,** which is **apqdr.** Since J ∩ (H ∨ K) and (J ∩ H) ∨ (J ∩ K) have the same generator, they are the same subgroup.

In other words, for cyclic groups, if we take the set of all the subgroups, there is an algebraic structure that has two operations defined on it. Much later in our study of algebra, we will consider such structures, which are called distributive lattices.

Exercise 35: Make a lattice diagram for all of the cyclic groups up to Z_{32}.

Exercise 36: Make lattice diagrams for Z_{90}, Z_{80}, and Z_{100}. Find a generator for each subgroup.

Exercise 37: In Z_{72} find all the generators of the following subgroups: Z_{18}, Z_9, Z_{24}, and Z_6. Trace on the lattice diagram all their joins and intersections.

Lesson 11: The Group of Units of a Monoid

One advantage of using additive notation for the cyclic groups is that we have access to a second operation, namely multiplication. However, the elements of Z_n under multiplication do not form a group because not all the elements of Z_n have multiplicative inverses – and recall that **1** is the multiplicative identity. Consider the familiar Z_{12} as an example. Obviously **0** has no inverse; no element **k** has the property that $k \cdot 0 = 1$. There are other elements of Z_{12} that also do not have multiplicative inverses; for example, try to find a multiplicative inverse for **3** in Z_{12}.

Most of the cyclic groups Z_n under multiplication are not groups but monoids. However, it is possible to find a subset of these monoids which is in fact a group. We will prove this and introduce a bit of new terminology.

Definition 16: Let $< M, * >$ be a monoid. An element $m \in M$ which has an inverse with respect to $*$ is called a **unit** of M.

Theorem 16: Let $< M, * >$ be a monoid. Then the set of units of M under $*$ forms a group.

Proof: There is scarcely any proof necessary. Let G be the set of units of M. The identity in M is a unit by default. The operation $*$ is associative on the elements of G because they are all also elements of M and $*$ is associative on the elements of M. Finally the elements of G have inverses with respect to $*$ by definition of unit.

Thus we look at the group of units of $< Z_n, \cdot >$. Since this is a completely different algebraic structure from the monoid $< Z_n, \cdot >$ we need a new symbol for it. We will denote the group of units of Z_n under multiplication by U_n. Now in U_n we can, for the first time, consider some equations worthy of the name, because we can continue using + that is inherited from Z_n. To consider such equations either in the monoid $< Z_n, \cdot >$ or in the group $< U_n, \cdot >$, we will need a different notation in order to make it clear that we are not doing algebra in the integers. Further, since we are using numerals to represent the elements of Z_n it cam be somewhat jarring to write $11 \cdot 5 = 7$ in U_{12} since that is obviously not true of ordinary integers, and it is not true in U_{13} where $11 \cdot 5 = 3$. Thus we will use the equivalence sign, $\equiv$, in place of =, and we will indicate which group of units we are doing the arithmetic in by writing (mod n) after the equation. Thus we will write $11 \cdot 5 \equiv 7$ (mod 12) and all will be clear. We will use this notation for either computing in the monoid Z_n or in the group U_n. An equation written in this form for elements of Z_n or U_n, is called a *congruence equation*. We'll state this more formally later.

Now we must be careful to remember that U_n does not include all the elements of Z_n but only those with multiplicative inverses. For example **3** does not have a multiplicative inverse in Z_{12}. To see this, we can compute the product of **3** with every other element of Z_{12} and see that **1** is never the answer.

$3 \cdot 0 \equiv 0$ (mod 12)	$3 \cdot 1 \equiv 3$ (mod 12)	$3 \cdot 2 \equiv 6$ (mod 12)	$3 \cdot 3 \equiv 9$ (mod 12)
$3 \cdot 4 \equiv 0$ (mod 12)	$3 \cdot 5 \equiv 3$ (mod 12)	$3 \cdot 6 \equiv 6$ (mod 12)	$3 \cdot 7 \equiv 9$ (mod 12)
$3 \cdot 8 \equiv 0$ (mod 12)	$3 \cdot 9 \equiv 3$ (mod 12)	$3 \cdot 10 \equiv 6$ (mod 12)	$3 \cdot 11 \equiv 9$ (mod 12)

Simply trying all the possible products is all very well in something with as few elements of Z_n but we will need a much better method to determine which elements of Z_n are elements of U_n and which are not. The next theorem answers the question completely.

Theorem 17: The element $m \in Z_n$ has a multiplicative inverse iff g.c.d. (m,n) = 1.

Proof: $\Rightarrow$) Suppose $m \in Z_n$ has a multiplicative inverse. Then there is some element $k \in Z_n$ such that $m \cdot k \equiv 1$ (mod n). Translating this congruence equation into an ordinary equation of integers, it says mk = 1 + rn for some integer r. If g.c.d.(m,n) = q > 1, then q divides the LHS of the equation and also q divides rn. But if q divides rn then q can't divide rn + 1, a contradiction. Therefore we know that g.c.d.(m,n) = 1.

$\Leftarrow$) This part of the proof will be done in lesson 13.

Hence, for the prime p all the non-zero elements of $\mathbf{Z}_p$ form a group under multiplication. As our first example, then, let's investigate $\mathbf{U}_5$. The elements of $\mathbf{U}_5$ are the elements **1, 2, 3,** and **4** from $\mathbf{Z}_5$. Whenever we first investigate a new group, the first step should be to consider the subgroups generated by each element. **1** is the identity, so let's calculate $<2>$:

2·2 $\equiv \mathbf{2}^2$ (mod 5) $\equiv$ **4** (mod 5).

2·4 $\equiv \mathbf{2}^3$ (mod 5) $\equiv$ **8** (mod 5) $\equiv$ **3** (mod 5).

2·3 $\equiv \mathbf{2}^4 \equiv$ **6** (mod 5) $\equiv$ **1** (mod 5)

Thus **2** has generated the whole group $\mathbf{U}_5$. Hence $\mathbf{U}_5$ is a cyclic group of order 4 and is the same as $\mathbf{Z}_4$.

Now let's calculate $\mathbf{U}_{12}$. First list the elements that are relatively prime to **12** to see what we are working with. They are **1, 5, 7,** and **11**. Begin with the subgroup generated by **5** always remembering to compute everything mod 12:

5·5 $\equiv \mathbf{5}^2$ (mod 12) $\equiv$ **25** (mod 12) $\equiv$ **1** (mod 12). Thus **5** has order 2 in $\mathbf{U}_{12}$; it generates a copy of $\mathbf{Z}_2$

7·7 $\equiv \mathbf{7}^2$ (mod 12) $\equiv$ **49** (mod 12) $\equiv$ **1** (mod 12). So **7** is also order 2 in $\mathbf{U}_{12}$ and generates another copy of $\mathbf{Z}_2$

11·11 $\equiv \mathbf{11}^2$ (mod 12) $\equiv$ **121** (mod 12) $\equiv$ **1** (mod 12). Again **11** is order 2 and generates a third copy of $\mathbf{Z}_2$

This uses up all the elements of $\mathbf{U}_{12}$ and results in a completely new group for us. It is a group of order 4, but it is not cyclic since no single element generates the whole group. Instead, each element is order 2; there are three subgroups that are identical to $\mathbf{Z}_2$ unlike $\mathbf{Z}_4$ which has only one subgroup that is identical to $\mathbf{Z}_2$. This new group is called the *Klein 4-group* and is denoted by **V**. To the right are the lattice diagrams for both $\mathbf{Z}_4$ and **V** for comparison. Thus there are at least two distinct groups of order 4 with entirely different structures. This is our first glimpse into the meaning of the phrase "algebraic structure"; here a set of four elements can be organized algebraically in two distinct ways each of which forms a group.

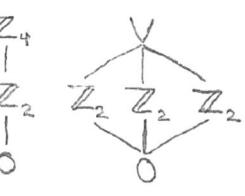

We will now introduce another way of exhibiting the structure of a group which is already familiar in the form of the "times tables" from elementary school. We write the elements of the group across the top of the table and also along the left hand side of the table. Then we write the product of the row element with the column element in the intersection of that row and column. Since groups in general will be non-commutative, we must have a rule about the construction of the table. The rule is that the element that names the row is the left element of the product, and the element that names the column is the right element of the product. Thus the result of the multiplication ab will be found at the intersection of row a and column b, but the result of the multiplication ba will be found at the intersection of row b and column a.

	1	a	b	c
1	1	a	b	c
a	a	1	c	b
b	b	c	1	a
c	c	b	a	1

	1	a	a^2	a^3
1	1	a	a^2	a^3
a	a	a^2	a^3	1
a^2	a^2	a^3	1	a
a^3	a^3	1	a	a^2

Whether the operation is represented as multiplication or addition or by *, we can form this table. It is called the *Cayley table* for the group. Above are the Cayley tables for both $\mathbf{Z}_4$ and **V** but I have renamed the elements for easier comparison. $\mathbf{Z}_4$ is written in multiplicative notation with the generator denoted by a; the other elements are naturally powers of a. The elements of **V** are each of order 2 and are named a, b, and c. The identities in both groups are denoted by 1.

It is easy to spot that these Cayley tables represent two different groups by the arrangement of the identity element within the table. It is also easy to tell from the table that the groups represented in them are Abelian because the tables are symmetric with respect to the diagonal; reflection through the diagonal reproduces the table because in an Abelian group the order of multiplication – that is the order of row and column – can be interchanged.

Exercise 38: Find the lattice diagram and Cayley table for U_{16}.

Exercise 39: Find the lattice diagram and Cayley table for U_{18}.

Exercise 40: Find the lattice diagram and Cayley table for U_{22}.

Felix Klein, for whom the Klein 4-group is named, was born in Dusseldorf in 1849. His father was secretary to the head of the Prussian government. Because his mother went into labor with him just as Rhinelander rebels began to attack the city, his family was not able to flee the city and he was born during the bombardment. He intended to be a physicist but soon changed to mathematics and received his degree at the University of Bonn. In 1875 he married Anne Hegel, the granddaughter of the famous philosopher. His rise in the mathematical world was rapid, but his health was always fragile. Due to the intensity with which he pursued research, he had a breakdown in 1882. After that, he devoted his energy mainly to teaching and to the reform of German education. He was the leading advocate for the equal admission of women into the universities. He is best known for his work putting Euclidean and non-Euclidean geometry on an equal footing. He developed a unified approach to geometry, which is still the standard viewpoint, making use of groups of transformations, though the definition of groups was not then fully assembled. He described the Klein 4-group in 1884. The Klein bottle also bears his name. His ill health forced him to retire in 1913, and he died in 1925.

"Thus, in a sense, mathematics has been most advanced by those who distinguished themselves by intuition rather than by rigorous proofs."

Arthur Cayley, for whom the Cayley table is named, was born in Richmond, Surrey, England in 1821, but lived in St. Petersburg, Russia, in his early childhood. He graduated Trinity College, Cambridge, in 1842 but could not obtain a university position at that time in England because he refused to accept Holy Orders. Instead, he became a lawyer for a livelihood while he pursued mathematics on the side. After 14 years practicing law, during which he wrote 250 mathematical papers, he was finally able to obtain a position at Cambridge, though it meant a sharp decrease in income. It was in 1854 that Cayley published a paper in which he gave the tables named for him. His study of permutations (which we take up in lesson 19) led him to formulate what later became the accepted definition of a group. He never fully accepted Klein's work on non-Euclidean geometry, believing it was circular reasoning. His mathematical life was extremely prolific, with over 900 papers published on nearly the whole range of mathematics. He died in 1895.

"As for everything else, so for a mathematical theory: beauty can be perceived but not explained."

Lesson 12: More About U_n

In order to practice using the ideas we have learned so far, we will determine the structure of some of the U_n that are a little more complicated. We will determine both the lattice diagram and the Cayley tables of these groups. Let's begin with U_{15}.

The numbers that are relatively prime to 15 are the following: **1, 2, 4, 7, 8, 11, 13,** and **14.** Thus U_{15} is a group of order 8. Unless this turns out to be cyclic we will have found a group completely new to us. As usual, begin by finding the subgroups generated by each of the elements. First $< \mathbf{2} >$.

$\mathbf{2 \cdot 2} \equiv \mathbf{2}^2$ (mod 15) $\equiv \mathbf{4}$ (mod 15).

$\mathbf{2 \cdot 4} \equiv \mathbf{2}^3$ (mod 15) $\equiv \mathbf{8}$ (mod 15).

$\mathbf{2 \cdot 8} \equiv \mathbf{2}^4$ (mod 15) $\equiv \mathbf{16}$ (mod 15) $\equiv \mathbf{1}$ (mod 15).

Thus **2** generates a cyclic subgroup of order 4, a copy of $\mathbf{Z_4}$. Since we know the internal working of $\mathbf{Z_4}$, we know that **8** also generates this same subgroup and that **4** generates the copy of $\mathbf{Z_2}$ that we know must be a subgroup of the $\mathbf{Z_4}$. Since **2, 4,** and **8** are accounted for we move on to $< \mathbf{7} >$.

$\mathbf{7 \cdot 7} \equiv \mathbf{7}^2$ (mod 15) $\equiv \mathbf{49}$ (mod 15) $\equiv \mathbf{4}$ (mod 15).

$\mathbf{7 \cdot 4} \equiv \mathbf{7}^3$ (mod 15) $\equiv \mathbf{28}$ (mod 15) $\equiv \mathbf{13}$ (mod 15).

$\mathbf{7 \cdot 13} \equiv \mathbf{7}^4$ (mod 15) $\equiv \mathbf{91}$ (mod 15) $\equiv \mathbf{1}$ (mod 15).

Hence **7** generates another copy of $\mathbf{Z_4}$, which is also generated by **13**. The copy of $\mathbf{Z_2}$ which we know must lie inside this second $\mathbf{Z_4}$ is the same one that we already found, the one generated by **4**. Thus we have found two copies of $\mathbf{Z_4}$ sharing a copy of $\mathbf{Z_2}$. The only elements not accounted for now are **11** and **14**. We will compute $< \mathbf{11} >$ next.

$\mathbf{11 \cdot 11} \equiv \mathbf{11}^2$ (mod 15) $\equiv \mathbf{121}$ (mod 15) $\equiv \mathbf{1}$ (mod 15).

This is then a new copy of $\mathbf{Z_2}$ not contained in any other subgroups we have found so far. The only element left to check is $< \mathbf{14} >$.

$\mathbf{14 \cdot 14} \equiv \mathbf{14}^2$ (mod 15) $\equiv \mathbf{196}$ (mod 15) $\equiv \mathbf{1}$ (mod 15).

Now we have a third copy of $\mathbf{Z_2}$ not related to the previous subgroups we have found, and we have found all the cyclic subgroups now.

But there is more to do because we now are aware of the existence of the Klein 4-group, $\mathbf{V}$. It consists of an identity element and three elements of order two. We have three elements of order two. That doesn't mean they form a subgroup together, but we should at least check it out. We will know they form a subgroup if they are closed under multiplication. It is easy to check. The set we are looking at are the elements of order 2 and is $\{\mathbf{1, 4, 11, 14}\}$. We compute:

$\mathbf{4 \cdot 11} \equiv \mathbf{44}$ (mod 15) $\equiv \mathbf{14}$ (mod 15).

$\mathbf{4 \cdot 14} \equiv \mathbf{56}$ (mod 15) $\equiv \mathbf{11}$ (mod 15).

$\mathbf{11 \cdot 14} \equiv \mathbf{154}$ (mod 15) $\equiv \mathbf{4}$ (mod 15).

We don't have to check anything else besides closure because **1** is the identity, and every element is its own inverse. Therefore these elements of order two combine to form a copy of $\mathbf{V}$. One of the three copies of $\mathbf{Z_2}$ in this $\mathbf{V}$ is shared with the two copies of $\mathbf{Z_4}$.

This allows us to draw the lattice diagram which is given below. I put the generators of each subgroup as additional help in understanding the group. The Cayley table is given beside the lattice diagram and you should check some of the products to satisfy yourself that they are correct. As far as we can tell, there are no other subgroups of U_{15} to find but we actually don't know this for sure. We know so little about groups and subgroups that there might be things hidden in U_{15} that we couldn't know anything about. There aren't but there could be. This is a brand new group to add to our list, only the second one that isn't cyclic.

As another example, let's determine the structure of U_{21}. As usual, find out what the elements are. The numbers that are relatively prime to 21 are: **1, 2, 4, 5, 8, 10, 11, 13, 16, 17, 19,** and **20,** so together

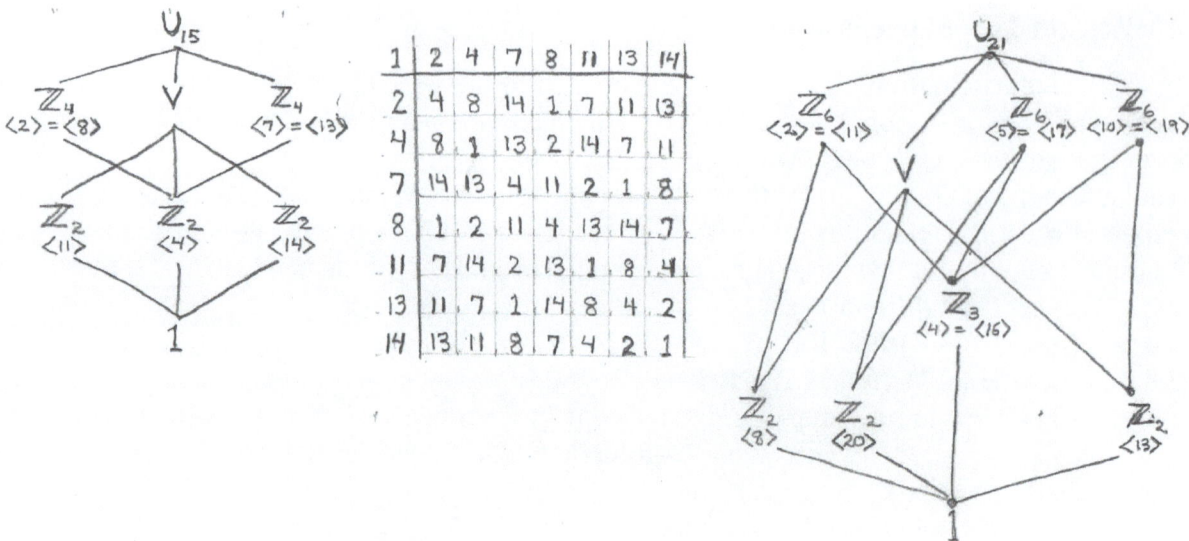

these form a group of order 12 under multiplication mod 21. As usual, we will first see what cyclic subgroups there might be. Begin with the smallest.

$2 \cdot 2 \equiv 2^2$ (mod 21) $\equiv 4$ (mod 21).
$4 \cdot 2 \equiv 2^3$ (mod 21) $\equiv 8$ (mod 21).
$8 \cdot 2 \equiv 2^4$ (mod 21) $\equiv 16$ (mod 21).
$16 \cdot 2 \equiv 2^5$ (mod 21) $\equiv 32$ (mod 21) $\equiv 11$ (mod 21).
$11 \cdot 2 \equiv 2^6$ (mod 21) $\equiv 22$ (mod 21) $\equiv 1$ (mod 21).

So $<2>$ is a copy of Z_6. Its other element of order six is **11**; its two elements of order three are **4** and **16**; and its element of order two is **8**. Six elements are left to check, the next smallest one being **5**.

$5 \cdot 5 \equiv 5^2$ (mod 21) $\equiv 25$ (mod 21) $\equiv 4$ (mod 21).
$5 \cdot 4 \equiv 5^3$ (mod 21) $\equiv 20$ (mod 21) $\equiv -1$ (mod 21).
$5 \cdot (-1) \equiv 5^4$ (mod 21) $\equiv -5$ (mod 21) $\equiv 16$ (mod 21).
$5 \cdot 16 \equiv 5^5$ (mod 21) $\equiv 80$ (mod 21) $\equiv 17$ (mod 21).
$5 \cdot 17 \equiv 5^6$ (mod 21) $\equiv 85$ (mod 21) $\equiv 1$ (mod 21).

Thus, we have another copy of Z_6. The elements of order six are **5** and **17**; the elements of order three are **4** and **16** – so we have a shared copy of Z_3 between the two Z_6's; and the element of order two is **20**, giving a second copy of Z_2. There are still three elements left to check, and we will look at **10** next.

$10 \cdot 10 \equiv 10^2$ (mod 21) $\equiv 100$ (mod 21) $\equiv 16$ (mod 21).
$10 \cdot 16 \equiv 10^3$ (mod 21) $\equiv 160$ (mod 21) $\equiv 13$ (mod 21)
$10 \cdot 13 \equiv 10^4$ (mod 21) $\equiv 130$ (mod 21) $\equiv 4$ (mod 21).
$10 \cdot 4 \equiv 10^5$ (mod 21) $\equiv 40$ (mod 21) $\equiv 19$ (mod 21).
$10 \cdot 19 \equiv 10^4$ (mod 21) $\equiv 190$ (mod 21) $\equiv 1$ (mod 21).

This is a third copy of Z_6. The elements of order six are **10** and **19**; the elements of order three are **16** and **4**, the same copy of Z_3 we already had; the element of order two is **13**, giving a third copy of Z_2.

The last thing to check is whether there is a copy of **V**, which may be present since we have three elements of order two, **8**, **13**, and **20**. Check by multiplying any two to see if the third is the result:

$8 \cdot 20 \equiv 160$ (mod 21) $\equiv 13$ (mod 21).

Hence the three copies of Z_2 together form a copy of **V**. The lattice diagram is given below. Again we have found a new non-cyclic group to add to our list. The Cayley table for U_{21} is left as an exercise.

Exercise 41: Construct the lattice diagram and Cayley table for U_{28}.

Exercise 42: Construct the lattice diagram and Cayley table for U_{27}.

Exercise 43: Construct the lattice diagram and Cayley table for U_{36}.

Lesson 13: The Euclidean Algorithm

We know that U_n under multiplication is a group, so we know that each element of U_n has a multiplicative inverse. Finding that inverse in a particular U_n is a daunting task, however, and if we don't know how to find inverses then a whole range of algebraic tools will be denied to us. For example, what is the inverse of **43** in U_{96}? Expressed as a congruence, we are asking for the value of x that solves the equation **43**x ≡ **1** (mod 96). How is such a number to be found?

Fortunately, one of the first mathematicians in history, Euclid, provided us with a method of calculating these inverses. We do not know if it was Euclid himself who discovered the method, but the earliest record of this result occurs in Euclid's book and so he gets the credit. The purpose of the Euclidean algorithm is to find the greatest common divisor of two integers and it is quite efficient at doing it. But it turns out the greatest common divisor is exactly what we need in order to find the inverse of an element in U_n.

In order to calculate g.c.d.(n,m), looking for an integer that will divide them both, it makes some sense to divide the larger by the smaller. If the division has no remainder, then the smaller of the two is the greatest common divisor. Otherwise the division will give us a quotient and a remainder. The Euclidean algorithm has us divide the remainder into the divisor; this results in a new quotient and a new remainder smaller than the first one. We will continue dividing each succeeding remainder into the preceding divisor, which was the previous remainder; and since each division will yield smaller positive remainders, we must eventually reach a point at which we get no remainder. This is the focus of the next theorem.

Theorem 18: The last non-zero remainder given by the Euclidean algorithm for the integers n and m equals g.c.d. (n,m).

Proof: Assume n > m. We get a series of equations as we execute this process. Since we will only be interested in the remainders, not the quotients, we will put the original two numbers and all the remainders in boldface.

$\mathbf{n} = \mathbf{m} \cdot q_1 + \mathbf{r_1}$ equation 1
$\mathbf{m} = \mathbf{r_1} \cdot q_2 + \mathbf{r_2}$ equation 2
$\mathbf{r_1} = \mathbf{r_2} \cdot q_3 + \mathbf{r_3}$ equation 3
⋮
$\mathbf{r_{n-3}} = \mathbf{r_{n-2}} \cdot q_{n-1} + \mathbf{r_{n-1}}$ equation n–1
$\mathbf{r_{n-2}} = \mathbf{r_{n-1}} \cdot q_n + \mathbf{r_n}$ equation n
$\mathbf{r_{n-1}} = \mathbf{r_n} \cdot q_{n+1}$ equation n+1

We wish to show that $\mathbf{r_n}$ is the greatest common divisor of $\mathbf{n}$ and $\mathbf{m}$. Consider the equations in reverse. From equation n+1 we see that $\mathbf{r_n} \mid \mathbf{r_{n-1}}$. Therefore $\mathbf{r_n}$ can be factored out of the RHS of equation n and hence $\mathbf{r_n} \mid \mathbf{r_{n-2}}$. Since $\mathbf{r_n}$ divides both $\mathbf{r_{n-1}}$ and $\mathbf{r_{n-2}}$ we know that $\mathbf{r_n}$ can be factored out of the RHS of equation n–1 and hence divides $\mathbf{r_{n-3}}$. We can follow this procedure all the way to equation 2 to conclude that $\mathbf{r_n} \mid \mathbf{m}$ and then back to equation 1 to conclude $\mathbf{r_n} \mid \mathbf{n}$. Thus we have shown $\mathbf{r_n}$ is a common divisor but not that it is the greatest one.

To this end, suppose there is a larger common divisor of $\mathbf{m}$ and $\mathbf{n}$, say $t > \mathbf{r_n}$. Then equation 1 tells us that $t \mid \mathbf{r_1}$. Moving on to equation 2, since $t \mid \mathbf{r_1}$ and $t \mid \mathbf{m}$, then $t \mid \mathbf{r_2}$. We can follow this trail back down the sequence of equations and conclude that $t \mid \mathbf{r_n}$, which contradicts that $t > \mathbf{r_n}$. Hence $\mathbf{r_n}$ = g.c.d. (**n**,**m**)

To see the method in action, let's compute g.c.d.(248, 1296).

    1296 = 5·**248** + **56**   (eq. 1)
    248 = 4·**56** + **24**   (eq. 2)

$$\mathbf{56} = 2 \cdot \mathbf{24} + \mathbf{8} \quad \text{(eq. 3)}$$
$$\mathbf{24} = 3 \cdot \mathbf{8}$$

As the last non-zero remainder, 8 = g.c.d.(248, 1296). The Euclidean algorithm is a very efficient method for computing greatest common divisors and works well even with very large numbers.

But the Euclidean algorithm gives us and unexpected dividend (pun intended), which is more valuable to our present concern. Once the greatest common divisor has been calculated, we can use the equations in reverse to express the g.c.d.(n,m) as a linear combination of n and m; in other words, it shows us how to find two numbers x and y such that, if d = g.c.d.(n,m) then d = x·n + y·m.

Referring to the equations in the proof of theorem 18, first solve equation n for $\mathbf{r_n}$ and label this equation n'. Then solve equation n–1 for $\mathbf{r_{n-2}}$ and substitute the resulting expression into the equation n'. Once this is simplified, it becomes the new equation n'. Continue solving each equation for its remainder, substituting into equation n', and simplifying the result.

$$\mathbf{r_{n-2}} = \mathbf{r_{n-1}} \cdot q_n + \mathbf{r_n} \quad \rightarrow \quad \mathbf{r_n} = \mathbf{r_{n-2}} - \mathbf{r_{n-1}} \cdot q_n \quad \text{(equation n')}$$
$$\mathbf{r_{n-3}} = \mathbf{r_{n-2}} \cdot q_{n-1} + \mathbf{r_{n-1}} \quad \rightarrow \quad \mathbf{r_{n-1}} = \mathbf{r_{n-3}} - \mathbf{r_{n-2}} \cdot q_{n-1}$$

now substitute into equation n': $\mathbf{r_n} = \mathbf{r_{n-2}} - \mathbf{r_{n-1}} \cdot q_n = \mathbf{r_{n-2}} - (\mathbf{r_{n-3}} - \mathbf{r_{n-2}} \cdot q_{n-1}) \cdot q_n$
$$= \mathbf{r_{n-2}} - \mathbf{r_{n-3}} \cdot q_n + \mathbf{r_{n-2}} \cdot q_{n-1} \cdot q_n = \mathbf{r_{n-2}} \cdot (1 + q_{n-1} \cdot q_n) - \mathbf{r_{n-3}} \cdot q_n$$

Eventually when we arrive at equation 1, $\mathbf{r_n}$ will be expressed in terms of **m** and **n** only. Let's consider the specific example above.

$$\mathbf{56} = 2 \cdot \mathbf{24} + \mathbf{8} \quad \rightarrow \quad \mathbf{8} = \mathbf{56} - 2 \cdot \mathbf{24} \quad \text{(from eq. 3)}$$
$$\mathbf{248} = 4 \cdot \mathbf{56} + \mathbf{24} \quad \rightarrow \quad \mathbf{24} = \mathbf{248} - 4 \cdot \mathbf{56} \quad \text{(from eq. 2)}$$

Hence $\mathbf{8} = \mathbf{56} - 2 \cdot \mathbf{24} = \mathbf{56} - 2 \cdot (\mathbf{248} - 4 \cdot \mathbf{56}) = \mathbf{56} - 2 \cdot \mathbf{248} + 8 \cdot \mathbf{56}$ (substituting into eq. 3)
$$\mathbf{8} = 9 \cdot \mathbf{56} - 2 \cdot \mathbf{248} \quad \text{(equation 3')}$$
$$\mathbf{1296} = 5 \cdot \mathbf{248} + \mathbf{56} \quad \rightarrow \quad \mathbf{56} = \mathbf{1296} - 5 \cdot \mathbf{248} \quad \text{(from eq. 1)}$$

Hence $\mathbf{8} = 9 \cdot \mathbf{56} - 2 \cdot \mathbf{248} = 9 \cdot (\mathbf{1296} - 5 \cdot \mathbf{248}) - 2 \cdot \mathbf{248}$ (substituting into eq. 3')
$$\mathbf{8} = 9 \cdot \mathbf{1296} - 45 \cdot \mathbf{248} - 2 \cdot \mathbf{248}$$
$$\mathbf{8} = 9 \cdot \mathbf{1296} - 47 \cdot \mathbf{248}$$

It is using the Euclidean algorithm in reverse that allows us to compute inverses in $\mathbf{U_n}$. As we noted at the beginning of this lesson, to find the inverse of **m** in $\mathbf{U_n}$ means to find the element **k** in $\mathbf{U_n}$ such that $\mathbf{m \cdot k} \equiv \mathbf{1}$ (mod n). Since **m** is in $\mathbf{U_n}$, we know g.c.d. (m, n) = 1, so that applying the Euclidean algorithm to m and n would yield 1 as the last non-zero remainder, and the equations that we would generate by using the algorithm can be used in reverse to express 1 as a linear combination of m and n; that is, we can find x and k such that $\mathbf{m \cdot k} + x \cdot n = \mathbf{1}$, which is the same as $\mathbf{m \cdot k} = \mathbf{1} - x \cdot n$, which is the same as $\mathbf{m \cdot k} \equiv \mathbf{1}$ (mod n). The value of x is of no consequence since we are doing the arithmetic mod n and any multiple of n is congruent to 0.

Thus, to find the inverse of **m** in $\mathbf{U_n}$ simply use the Euclidean algorithm to find g.c.d. (m, n), which is always 1; then use the resulting equations in reverse order to express 1 as a linear combination of n and m; the coefficient of m is its multiplicative inverse mod n.

In the example above, we found that g.c.d.(248, 1296) = 8. Divide both 248 and 1296 by 8 and we have relatively prime numbers, 31 and 162 and dividing 8 into the linear combination we derived above we get $\mathbf{1} = 9 \cdot \mathbf{162} - 47 \cdot \mathbf{31}$. This can be interpreted in either of two ways. In $\mathbf{U_{31}}$ the multiplicative inverse of **162** is **9**. $\mathbf{162} \equiv \mathbf{7}$ (mod 31) so **7** and **9** are inverses in $\mathbf{U_{31}}$. Or we could interpret $\mathbf{1} = 9 \cdot \mathbf{162} - 47 \cdot \mathbf{31}$ in $\mathbf{U_{162}}$. In this case the interpretation is more straight-forward: **47** is the multiplicative inverse of **31** in $\mathbf{U_{162}}$.

Exercise 44: Find the inverse of 77 in $\mathbf{U_{213}}$.
Exercise 45: Find the inverse of 3000 in $\mathbf{U_{7001}}$.
Exercise 46: Find the inverse of 97 in $\mathbf{U_{1110}}$.

Euclid, along with Pythagoras, is perhaps the most famous mathematician who ever lived and wrote what is without doubt the most successful math textbook ever written, still a standard today for studying geometry. However little is known about his life. We don't know where he was born or exactly when or any of the circumstances. We do know that he arrived at the famous Library of Alexandria around 322 b.c., about 10 years after it was founded by Alexander the Great. He wrote *The Elements* for the purpose of presenting all, or most, of the mathematics known in his day in a coherent orderly form, with carefully presented theorems and proofs. This book set the standard for all mathematics presentations ever since. The Euclidean Algorithm appeared in that book. Since none of the theorems are attributed to specific mathematicians, we really don't know if Euclid was the discoverer of the algorithm or not. The system of geometry he described in *The Elements* was named for him. Euclid died in about 270 b.c. but so little is known about the details of his life that some scholars doubt he even existed, that instead Euclid was a collection of people who worked together rather than an individual.

"What has been affirmed without proof, can also be denied without proof."

Lesson 14: Congruence Equations

We'll begin with the definition.

Definition 17: A **congruence equation** is any equation whose coefficients, variables and possible solutions are elements of $\mathbb{Z}_n$ or U_n.

This definition will be generalized in a much broader context eventually, but this will be quite adequate for now. We will be working exclusively with linear congruence equations, ones in which the variable is not raised to a non-trivial power. The easiest way of thinking about a congruence equation is to translate them into an equivalent equation in $\mathbb{Z}$ where they will seem more familiar. The translation is immediate from the way we defined $\mathbb{Z}_n$:

$$a\mathbf{x} \equiv \mathbf{m} \pmod{n} \text{ as an equation in } \mathbf{Z}_n \text{ or } \mathbf{U}_n \iff ax = m + kn \text{ as an equation in } \mathbb{Z}$$

We will continue to use boldface for any numeral that is considered as an element of $\mathbf{Z}_n$ rather than $\mathbb{Z}$.

Unless n is a prime, $\mathbf{Z}_n$ under multiplication mod n is a monoid and not a group. This lesson, then, is about doing algebra in a monoid. Since not all elements of a monoid have inverses, the cancellation laws do not work in general. We must be more clever, then, in solving some of these equations and for that purpose the Euclidean algorithm will be a basic tool. First we need to see how far we can extend the cancellation laws in $\mathbf{Z}_n$ under multiplication.

Theorem 19: If $c\mathbf{a} \equiv c\mathbf{b} \pmod{n}$ then $\mathbf{a} \equiv \mathbf{b} \pmod{n/d}$ where $d = \text{g.c.d.}(c, n)$.

Proof: First translate $c\mathbf{a} \equiv c\mathbf{b} \pmod{n}$ into $ca = cb + kn$. Let $d = \text{g.c.d.}(c, n)$, so that we know that $c = dc'$, $n = dn'$, and c' and n' are relatively prime. We can rewrite the equation as follows:

$$c\mathbf{a} \equiv c\mathbf{b} \pmod{n} \iff dc'a = dc'b + kdn'$$

$$\iff c'a = c'b + kn' \text{ (because the cancellation laws hold in } \mathbb{Z})$$

$$\iff c'\mathbf{a} \equiv c'\mathbf{b} \pmod{n'}$$

$$\iff c'\mathbf{a} \equiv c'\mathbf{b} \pmod{n/d}$$

Now we use what we know about $\mathbf{U}_n$ as a sub-monoid of $\mathbf{Z}_n$. Since c' is relatively prime to n/d, it has a multiplicative inverse in $\mathbf{Z}_n$. If we multiply both sides by $(\mathbf{c'})^{-1}$ then we get the required result: $\mathbf{a} \equiv \mathbf{b} \pmod{n/d}$.

In short, we have the rather unexpected result that we can cancel a common factor on both sides of the equation but only if we change the monoid we are solving the equation in, from $\mathbf{Z}_n$ to $\mathbf{Z}_{n/d}$. Of course, if c is relatively prime to n, then it has a multiplicative inverse and we can simply cancel it without changing the monoid. This allows us to classify linear congruences into ones that have a solution and ones that don't.

Theorem 20: Let $d = \text{g.c.d.}(n,a)$. The equation $a\mathbf{x} \equiv \mathbf{m} \pmod{n}$ has *no solution* in $\mathbf{Z}_n$ if d does not divide m.

Proof: Translate the congruence equation into an ordinary equation:

$$ax = m + kn, \text{ which is } ax - kn = m.$$

Since $d|a$ and $d|n$, d divides the LHS of the equation. But d does not divide the RHS of the equation, which is impossible.

Next we must consider the other possibility. What if d does divide m?

Theorem 21: Let $d = \text{g.c.d.}(n,a)$. If $d\,|\,\mathbf{m}$ then the equation $a\mathbf{x} \equiv \mathbf{m} \pmod{n}$ has exactly d solutions in $\mathbf{Z}_n$. If $\mathbf{s}$ is one such solution and $\mathbf{q} = n/d$, then $\mathbf{s} + \mathbf{q}$, $\mathbf{s} + 2\mathbf{q}$, ... , $\mathbf{s} + (\mathbf{n} - 1)\mathbf{q}$, all mod n, are also solutions of $a\mathbf{x} \equiv \mathbf{m} \pmod{n}$.

Proof: Since $d = \text{g.c.d.}(n,a)$ we can write $n = n'd$ and $a = a'd$ with $\text{g.c.d.}(n',a') = 1$; since $d\,|\,m$ we can write $m = m'd$, though m' may not be relatively prime to n' or a'. We can translate the congruence equation into $da'x = dm' + kdn'$ and cancel d on both sides of the equation, just as we did in the proof of theorem 19. This gives the ordinary equation a'x = m' + kn', which translates back to a congruence equation **a'x ≡ m'** (mod n'). Now since g.c.d.(n',a') =1, theorem 20 does not apply because 1|m'. Unfortunately, theorem 20 does not say that this equation has a solution. However, since **a'** is relatively prime to n', **a'** ∈ **U**$_{n'}$, a group. Therefore **a'** has an inverse, **(a')**$^{-1}$ which we now multiply on both sides of the congruence equation. This results in the solution **x ≡ (a')**$^{-1}$**(a')**$^{-1}$ (mod n'). This solution is not necessarily in **U**$_{n'}$ since m' may not be relatively prime to n' and thus not an element of **U**$_{n'}$ but it is certainly a solution in **Z**$_{n'}$.

Let's call this solution **s**. Then **a's ≡ m'** (mod n'). **a's** is congruent to **m'** plus any multiple of n' and since n is a multiple of n', **s** is also congruent to **m'** plus any multiple of n. Thus **a's** is also a solution of **a'x ≡ m'** (mod n). Multiply both sides of this equation by **d** to get **da's ≡ dm'** (mod n), so **s** is a solution of the original congruence, **ax ≡ m** (mod n).

Now consider **s + jq** (mod n) for integer j. Substitute this into the original congruence and get
$$a(s + jq) \pmod{n} \equiv as + jaq \pmod{n} \equiv m + (jan)/d \pmod{n} \equiv m + (ja')n \pmod{n}$$
$$\equiv m + \pmod{n}$$
Therefore **s + jq** is also a solution of **ax ≡ m** (mod n). These will be incongruent to each other as long as 0 ≤ j < d. To see why, suppose
$$s + jq \equiv s + kq \pmod{n} \iff jq \equiv kq \pmod{n} \iff jq - kq \equiv 0 \pmod{n}$$
$$\iff (j - k)q \equiv 0 \pmod{n} \iff (j - k)n/d \equiv 0 \pmod{n}$$
This is true iff d | j − k, which is impossible if both j and k are less than d. Hence there are exactly d distinct solutions incongruent mod n.

For example, let's solve **9x ≡ 15** (mod 77). The g.c.d.(77, 9) = 1, so theorem 20 tells us that there is a solution. To find the multiplicative inverse of **9** use the Euclidean algorithm to find the g.c.d.(9,77), even though we know it is **1**.
$$77 = 8 \cdot 9 + 5 \implies 9 = 1 \cdot 5 + 4 \implies 5 = 1 \cdot 4 + 1 \implies 1 = 5 - 1 \cdot 4$$
Since **4 = 9 − 1·5** we have **1 = 5 − 1·(9 − 1·5) = 2·5 − 1·9**
Since **5 = 77 − 8·9** we have **1 = 2·(77 − 8·9) − 1·9 = 2·77 − 17·9**
Hence **1 ≡ − 17·9** (mod 77) **≡ 60·9** (mod 77)
Thus **60** is the multiplicative inverse of **9**. Multiply both sides of the original congruence by **60**:
  9x ≡ 15 (mod 77) ⟹ **60·9x ≡ 60·15** (mod 77) ⟹ **x ≡ 900** (mod 77) ≡ **53** (mod 77)
So **53** is the solution, the only solution.

As another example, we will find all the solutions of **12x ≡ 40** (mod 196). First we need to notice g.c.d.(12,196) = 4 and since 4 divides 40 we know that solutions exist and there are four of them by theorem 21. Next we reduce the equation to a simpler one by dividing both sides by 4 to get **3x ≡ 10** (mod 49), which has a unique solution. Using the Euclidean algorithm on 3 and 49, to get the inverse of **3**. We get: 49 = 16·3 + 1. So, after only one step we get
  1 ≡ −16·3 (mod 49) **≡ 33·3** (mod 49)
33 is the multiplicative inverse of **3** so multiply both sides by **33** and we get
  3x ≡ 10 (mod 49) ⟹ **33·3x ≡ 33·10** (mod 49) ⟹ **x ≡ 330** (mod 49) ≡ **36** (mod 49)
Hence x_0 ≡ **36** (mod 49) is a solution of **3x ≡ 10** (mod 49). Therefore x_0 ≡ **36** (mod 196) is a solution of **12x ≡ 40** (mod 196). The other three solutions of **12x ≡ 40** (mod 196) are now easily found:
  36 + 49 ≡ 85, 36 + 2·49 ≡ 134, and 36 + 3·49 ≡ 183.

Exercise 47: Solve **140x ≡ 133** (mod 301).
Exercise 48: Solve **35x ≡ 11** (mod 221).
Exercise 49: Solve **21x ≡ 42** (mod 54).
Exercise 50: Solve **−56x ≡ 7** (mod 154).

Lesson 15: Euler's Phi Function

We now know how to do some basic algebra in $\mathbf{U_n}$ and in $\mathbf{Z_n}$. Much of this algebra depends on knowing the numbers relatively prime to n and naturally, if n is very large, it can be quite difficult to list them all. We can, however, compute how many numbers smaller than n are relatively prime to it. Since the only elements of $\mathbf{U_n}$ are these relatively prime numbers, knowing how many there are also tells us the order of $\mathbf{U_n}$.

The name of the formula that computes the number of numbers relatively prime to n is the *Euler phi function*, denoted by $\varphi(n)$. By definition $\varphi(n)$ equals the number of positive integers less than n that are relatively prime to n, including the number **1**. We can write symbolically $|\mathbf{U_n}| = \varphi(n)$. We will now prove two theorems which, together, allow us to calculate $\varphi(n)$ for any integer n. Since integers are composed of prime numbers, it is natural that we should compute those values first.

Theorem 22: Let p be a prime. Then $\varphi(p^k) = p^{k-1}(p-1) = p^k - p^{k-1}$.

Proof: First this is immediate for the case k = 1. For the prime p, by definition every integer less than p is relatively prime to p. Thus $\varphi(p) = p - 1 = p^0(p-1)$.

Now if k > 1, it is clear that the only divisors of p^k are powers of p where the power is less than k. In other words, the only numbers not relatively prime to p^k are the ones that have p as a factor - that is, the multiples of p. How many multiples of p are there that are less than p^k? Clearly, $p^{k-1} - 1$ of them. Therefore, not counting p^k itself, there are
$$p^k - 1 - (p^{k-1} - 1) = p^k - p^{k-1} = p^{k-1}(p-1)$$
numbers that are relatively prime to p^k.

With the following theorem, we get everything we want. The proof of theorem 19 is an interesting one in its own right. The following is a standard proof for the next result.

Theorem 23: Let m, n be any natural numbers greater than 1 and relatively prime to one another. Then $\varphi(mn) = \varphi(m)\varphi(n)$.

Proof: First note that numbers relatively prime to mn must be simultaneously relatively prime to m and relatively prime to n. To prove the theorem, we will first arrange the integers from 1 up to mn in a rectangular array. Without loss of generality we will assume m<n:

1	2	3	.	.	.	m–1	m
m+1	m+2	m+3	.	.	.	2m–1	2m
2m+1	2m+2	2m+3	.	.	.	3m–1	3m
.	.	.	.	.	.	.	.
.	.	.	.	.	.	.	.
(n–2)m+1	(n–2)m+2	(n–2)m+3	.	.	.	(n–1)m–1	(n–1)m
(n–1)m+1	(n–1)m+2	(n–1)m+3	.	.	.	nm–1	nm

Now notice that for any given column, every number appearing in that column is congruent mod m to every other number in that column. So on the first row, if the number is relatively prime to m then every number in that column is relatively prime to m as well. There are $\varphi(m)$ such columns, each including n numbers. We can eliminate the other columns and that leaves us with $\varphi(m) \cdot n$ numbers that are less than mn and relatively prime to m.

Next, how many numbers in a given column are relatively prime to n? I claim that in any given column no two numbers can be congruent mod n. Consider the form of the numbers in, say, the rth column. It consists of the numbers r, m + r, 2m + r, …, (n – 1)m + r. *Suppose* two of these numbers were congruent mod n, say jm + r ≡ km + r (mod n) and assume j > k. Then we can subtract r from both sides to get jm ≡ km (mod n). This translates to

$$jm = km + qn \text{ for some integer q}$$
$$jm - km = qn \text{ for some integer q}$$
$$(j - k)m = qn.$$

Now n does not divided m, but n must divide the LHS of the equation, so $n \mid (j - k)$. This, however, is impossible because j and k are both less than n and therefore $j - k < n$. Thus our supposition must have been wrong and no two numbers in a given column are congruent. Since there are exactly n of them, each column contains one number congruent to each of $0, 1, 2, 3, \ldots, n - 1$. How many of these numbers in a single column will be relatively prime to n? Clearly $\varphi(n)$ of them, in each of the $\varphi(m)$ columns. Thus, the total number of integers relatively prime to mn is $\varphi(m)\varphi(n)$, and the theorem is proven.

Combining these two theorems gives us a way of computing $\varphi(n)$. First find the prime factorization of n (admittedly, this may be difficult to do); then use theorem 19 to reduce $\varphi(n)$ to a product of factors of the form $\varphi(p^k)$; finally use theorem 18 to compute each factor. For example, let's compute $\varphi(4050)$:

$$4050 = 2 \cdot 3^4 \cdot 5^2 \Rightarrow \varphi(4050) = \varphi(2)\cdot\varphi(3^4)\cdot\varphi(5^2) = 1\cdot 3^3 (3 - 1)\cdot 5(5 - 1) = 27\cdot 2\cdot 5\cdot 4 = 1080$$

Therefore we know $|U_{4050}| = 1080$. Eventually we will be able to identify the group U_{4050} by a name that informs us of its structure in some detail but we'll need more terminology and better methods.

There are two more theorems we can prove about the Euler phi function that will prove useful in the next lesson.

Theorem 24: Let $\{d_1, d_2, \ldots, d_m\}$ be the set of all the divisors of n. Then $\sum_{i=1}^{m} \varphi(d_i) = n$.

Proof: We will parition the integers from 1 to n into the following m sets:
$$S_i = \{k \mid 1 \leq k \leq n \text{ and g.c.d.}(n.k) = d_i\}$$
Now k is in S_i iff g.c.d.(k,n) = d_i which is true iff g.c.d.$(k/d_i, n/d_i) = 1$ – that is, iff k/d_i is relatively prime to n/d_i. Thus there are exactly $\varphi(n/d_i)$ integers in S_i. Since each integer between 1 and n is in one of these sets, we know $n = \sum_{i=1}^{m} |S_i| = \sum_{i=1}^{m} \varphi(n/d_i)$. But as d_i ranges over all the divisors of n, so does n/d_i. Therefore $n = \sum_{i=1}^{m} \varphi(d_i)$ as claimed.

The next theorem is one that can be proven as a mere corollary to later theorems which we cannot do at the moment. Since we need the result now, we will settle for a longer proof.

Theorem 25 (Euler's Theorem): Let $m \in U_n$, then $m^{\varphi(n)} \equiv 1 \pmod{n}$.

Proof: Let $k = \varphi(n)$ and let $a_1, a_2, \ldots, a_k$ be a list of the elements of U_n in some order. Clearly, m is on this list. Then $ma_1, ma_2, \ldots, ma_k$, reduced mod n, is still a list of the elements of U_n though in some other order. We can let
$$ma_1 \equiv a_1' \pmod{n}, \quad ma_2 \equiv a_2' \pmod{n}, \quad \ldots, \quad ma_k \equiv a_k' \pmod{n}.$$
Now multiply this sequence of k congruences together; the product of the LHS must be congruent to the product of the RHS:
$$(ma_1)(ma_2)\ldots(ma_k) \equiv a_1'a_2'\ldots a_k' \pmod{n} \Rightarrow m^k(a_1 a_2 \ldots a_k) \equiv a_1'a_2'\ldots a_k' \pmod{n}$$
Since the a_i' are the same numbers as the a_i though in a different order, $a_1 a_2 \ldots a_k = a_1'a_2'\ldots a_k'$. They are all relatively prime to n and so their product is also relatively prime to n. Then the product has a multiplicative inverse and we can multiply both sides by this inverse. This gives $m^k \equiv 1 \pmod{n}$. Since $k = \varphi(n)$, this is what we set out to prove.

Note that this does not say that the order of the element m is $\varphi(n)$. It only says that the order of m must be a divisor of $\varphi(n)$.

Exercise 51: Find $|U_{192080}|$.

Exercise 52: Find $|U_{1,328,096}|$.

Leonhard Euler was born in 1707 in Basle, Switzerland, and was one of the most prolific mathematicians of all time. There is no area in the mathematics of his time that he did not transform. Originally, his father decided that he would become a minister. But his father was close friends with another prominent mathematician, Johann Bernoulli, who noticed the young Euler's talent for math and dissuaded his father from insisting on the ministry. Euler was a devout Christian all his life, but never had the passion for theology that he had for mathematics. He was offered a post at the university in St. Petersburg, Russia, in 1726. He had severe health problems in 1735, which resulted in the loss of sight in one eye by 1740. Due to political turmoil in Russia, he moved to Berlin in 1745, but returned to St. Petersburg in 1766. His health problems returned and by 1771 he became totally blind. This difficulty did not even slow down his mathematical research. His memory was so good he could perform extraordinarily complicated calculations in his head. He was productive mathematically until the day he died in 1783 of a brain hemorrhage. So great was his productivity that he could not publish papers as fast as he wrote them. His research, found in trunks, was published gradually for 50 years after his death. Euler's Theorem was published in 1763, and he introduced the Phi function also in 1763.

"Now I will have less distraction." (Euler, upon losing the use of his right eye; as quoted in *In Mathematical Circles* (1969) by H. Eves)

Lesson 16: Primitive Roots, Part 1

We now know the order of each of the U_n. We would like, of course, to completely determine the structure of these groups, but we only know the cyclic groups Z_n. In general, the groups U_n are more complicated and we are not ready yet to discuss them. However some of the U_n are cyclic; it is our immediate goal then to determine which of them are cyclic and which are not. Since we know the order of U_n is $\varphi(n)$, if we know U_n is cyclic then we know that $U_n = Z_{\varphi(n)}$. Then we would have all we need to understand these particular cyclic groups if we could find a generator. Our first goal is to determine which of the U_n are cyclic, and our second goal is to identify at least one generator.

This material was first studied by Gauss nearly a century before modern algebra crystallized. It forms a standard part of the material covered by an introductory number theory course, and much of the terminology arose in that setting. Thus we need to introduce a new definition.

Definition 18: In U_n any element that has order $\varphi(n)$ - that is, any element that generates U_n, - is said to be a **primitive root** of n.

Our first goal is to determine which integers n have primitive roots; that is, for which integers n is U_n cyclic. This is rather difficult to answer and will require several steps. In discovering which integers have a particular property and which do not, it is generally best to begin with the prime numbers, and that is what we do now. We will begin by proving even more than we need.

Theorem 26: Let p be a positive odd prime and assume $q \mid (p-1)$. Then there are $\varphi(q)$ elements of U_p which have order q.

Proof: Let S_q be the set of all the elements of U_p which have order q. We know that q must divide the order of U_p which is $\varphi(p) = (p-1)$. We also know that every element of U_p must be in one of the sets S_q, so if we count all the elements in each S_q and add them together the total must equal the number of elements in U_p which is $p - 1$:

$$\sum_q |S_q| = p - 1$$

By theorem 24, since q varies over all the divisors of $p - 1$, we also know that

$$\sum_q \varphi(q) = p - 1$$

Now let x be an element of order q; that is, $x \in S_q$. Then $< x >$ is the cyclic group Z_q and we know there are $\varphi(q)$ possible generators for $< x >$. This means that Z_q includes $\varphi(q)$ elements of order q. Therefore $|S_q| \geq \varphi(q)$ for each divisor q of $p - 1$. So the only way the two summations can be equal is if $|S_q| = \varphi(q)$. Thus there are $\varphi(q)$ elements of U_p which have order q.

Corollary 26a: Each odd prime p has a primitive root.

Proof: Clearly $(p-1) \mid (p-1)$. Therefore U_p, which has order $p - 1$, has at least one element of order $p - 1$, which is by definition a primitive root for p.

Corollary 26b: $U_p = Z_{p-1}$.

Proof: is immediate.

What integers do we examine after the primes? We will next determine which integers *do not* have primitive roots. As it turns out, most numbers do not. Two large classes of numbers don't.

Theorem 27: For $k \geq 3$, 2^k does not have a primitive root.

Proof: Since an integer is relatively prime to 2^k iff it is odd, we need only show that each odd number has an order different from $\varphi(2^k) = 2^{k-1}$. We will prove this by induction on k. First let $k = 3$. Then $2^3 = 8$ and $\varphi(8) = 4$, so $|U_8| = 4$. It is easy to check that $a^2 \equiv 1 \pmod{8}$ for each odd number less than 8 so each element of U_8 has order 2. Hence 8 has no primitive root. This also shows us, by the way, that U_8 is the Klein 4-group **V**.

Now assume the theorem is true for some particular integer m: assume 2^m does not have a primitive root. More specifically we will assume the order of an odd integer x which is less than 2^m is a divisor of $2^{m-2} = q$. Then $x^q = 1$ for all the elements in $U_{2^{\wedge}m}$ and in this way it has no single generator. This was exactly what we found for m = 3. Now we will use this to prove the same thing for m+1, that the order of any odd number in $U_{2^{\wedge}(m+1)}$ is a divisor of 2^{m-1} rather than 2^m as would be required for a generator.

Begin with what we assumed from the previous case m. Let $x \in U_{2^{\wedge}m}$ so that we know:
$$x^{2^{\wedge}(m-2)} \equiv 1 \pmod{2^m} \implies x^{2^{\wedge}(m-2)} = 1 + c \cdot 2^m$$
If we square both sides of the last equation, we get
$$x^{2^{\wedge}(m-1)} = 1 + 2c \cdot 2^m + (c \cdot 2^m)^2 = 1 + 2c \cdot 2^m + c^2 \cdot 2^{2m} = 1 + 2^{m+1}(c + c^2 \cdot 2^{m-1})$$
$$\equiv 1 \pmod{2^{m+1}}$$
Hence the theorem holds for the case m + 1, and by induction the theorem holds for all m.

Thus we know that U_{256}, for example, is a group of order 128 but it is not cyclic. What it is exactly we do not yet have any way of knowing, of course. We are next able to spot another large class of numbers which do not have primitive roots.

Theorem 28: Let m and n be relatively prime and both greater than 2. Then mn has no primitive root.

Proof: Let k be relatively prime to mn; that is, $k \in U_{mn}$. Then k is also relatively prime to both m and n as well. Note that both $\varphi(m)$ and $\varphi(n)$ are even numbers. Set $h = \text{l.c.m.}(\varphi(m), \varphi(n))$ and then set $d = \text{g.c.d.}(\varphi(m), \varphi(n))$. Then d is also even and $h = [\varphi(m) \cdot \varphi(n)]/d \leq \frac{1}{2} \cdot \varphi(mn)$.

Now by theorem 25, $k^{\varphi(m)} \equiv 1 \pmod{m}$. Raising both sides of this congruence to the power $\varphi(n)/d$ we get
$$k^{[\varphi(m) \cdot \varphi(n)]/d} \equiv 1 \pmod{m} \implies k^h \equiv 1 \pmod{m}.$$

We can use a parallel argument to conclude that $k^h \equiv 1 \pmod{n}$. So $k^h \equiv 1 \pmod{mn}$. Hence the order of k is at most $h \leq \frac{1}{2} \cdot \varphi(mn)$. Since k was an arbitrary element of U_{mn}, no one element of U_{mn} will generate the whole group and mn has no primitive root.

In the next lesson we will prove that all other integers do have primitive roots.

Exercise 53: Find all the generators of U_{13} and show how they generate the group explicitly.
Exercise 54: Find two generators of U_{19} and show how they generate the group explicitly.
Exercise 55: Find two generators of U_{31} and show how they generate the group explicitly.

Lesson 17: Primitive Roots, Part 2

The only numbers left that may have primitive roots are the numbers p^k and $2p^k$ where p is an odd prime. We now show that these numbers do indeed have primitive roots, but it will take a few more steps.

Theorem 29: Let p be an odd prime and r a primitive root of p. Then r or r + p or both are primitive roots of p^2.

Proof: First notice that the order of any element of $\mathbf{U}_{p^2}$ must divide $\varphi(p^2) = p(p-1)$. This means that if we can find an element whose order does not divide p – 1, then its order must equal either p or p(p – 1). Hence we will look for an element r such that $r^{p-1} \not\equiv 1 \pmod{p^2}$, and since r is a primitive root of p, its order can't equal p and so it must have order p(p – 1) and be a primitive root of p^2.

So let r be any primitive root of p. Then $r^{p-1} \equiv 1 \pmod p$. It may be already the case that for this choice $r^{p-1} \not\equiv 1 \pmod{p^2}$, and if so, as we just noticed, it is a primitive root and we are done. Therefore we will assume that $r^{p-1} \equiv 1 \pmod{p^2}$ and let s = r + p. Then use the Binomial Theorem to raise s to the p – 1 power:

$$s^{p-1} = (r+p)^{p-1} = r^{p-1} + (p-1)\cdot r^{p-2}\cdot p + \tfrac{1}{2}\cdot(p-1)(p-2)r^{p-3}\cdot p^2 + \ldots + p^{p-1}$$

$$s^{p-1} \equiv (r+p)^{p-1} \pmod{p^2} \equiv 1 + (p^2-p)\cdot r^{p-2} + \tfrac{1}{2}\cdot(p-1)(p-2)r^{p-3}\cdot p^2 + \ldots + p^{p-1} \pmod{p^2}$$

When expanded, all but the first two terms will involve powers of p higher than 2 and so are congruent to 0 $\pmod{p^2}$. Hence

$$s^{p-1} \equiv 1 - p\cdot r^{p-2} \pmod{p^2}.$$

Now because p is relatively prime to r, $p \nmid r^{p-2}$ and therefore $p^2 \nmid p\cdot r^{p-2}$. Thus

$$s^{p-1} \not\equiv 1 \pmod{p^2}.$$

Hence s must have order p(p – 1) and is a primitive root of p^2.

As an example, let's find a primitive root of 49. By theorem 29 we can reduce this problem to the easier problem of finding a primitive root of 7. The first one we try, 2, is easy to check:

$$2^3 \equiv 8 \pmod 7 \equiv 1 \pmod 7 \text{ so the order of 2 is 3, not 6.}$$

But if 2 has order 3, this suggests that 4 will have order 3 as well, so lets try 3 next:

$$3^2 \equiv 9 \pmod 7 \equiv 2 \pmod 7$$
$$3^3 \equiv 3\cdot 2 \pmod 7 \equiv 6 \pmod 7 \equiv -1 \pmod 7$$

This implies that the order of 3 is 6 and so 3 is a primitive root of 7. By theorem 29, either 3 or 10 is a primitive root of 49; naturally we try 3 first. Since $\varphi(49) = 42$, we need to compute the powers of 3 mod 49 up to the 42nd power, but we don't need them all. We just need to be sure that 3 is not congruent to 1 for some power smaller than 42 that divides 42; that is, for powers 2, 3, 6, 7, 14, or 21. Clearly 3^2 and 3^3 are not congruent to 1 (mod 49). We will now compute 3^6, 3^7, 3^{14}, and 3^{21} if necessary.

$$3^6 \equiv 3^3\cdot 3^3 \equiv 27\cdot 27 \equiv 729 \equiv 43 \equiv -6 \pmod{49}$$
$$3^7 \equiv 3^6\cdot 3 \equiv -6\cdot 3 \equiv -18 \equiv 31 \pmod{49}$$
$$3^{14} \equiv 3^7\cdot 3^7 \equiv 31\cdot 31 \equiv 961 \equiv 30 \pmod{49}$$
$$3^{21} \equiv 3^7\cdot 3^{14} \equiv 31\cdot 30 \equiv 930 \equiv 48 \pmod{49} \equiv -1 \pmod{49}$$

Therefore $3^{42} \equiv 1 \pmod{49}$ and no smaller power of 3 can be congruent to 1. Hence 3 is a primitive root of 49 and is a generator of $\mathbf{U}_{49} = \mathbf{Z}_{42}$. If 3 had failed to be a primitive root, then we would have done the same checking with 10, but theorem 29 would have guaranteed that 10 would be a primitive root if 3 had failed and checking it would not have been necessary. We could display $\mathbf{U}_{49} = <3>$ by computing all the powers of 3 mod 49 and showing how every element of $\mathbf{U}_{49}$ is produced in turn.

We will now show that p^k has a primitive root for all positive values of k. The proof is very similar to the proof of theorem 25.

Theorem 30: Let p be an odd prime. Then p^k has a primitive root.

Proof: Begin with r, a primitive root of the prime p with the additional condition, as in the proof of theorem 29, that $r^{p-1} \not\equiv 1 \pmod{p^2}$. We will proceed by induction on k to prove the more general congruence
$$r^{(p-1)p^{k-2}} \not\equiv 1 \pmod{p^k}.$$
The base case, $k = 2$, is true by assumption. We will therefore assume the theorem is true for some $k > 2$, and prove it for $k + 1$. That is,

we assume $\quad r^{(p-1)p^{k-2}} \not\equiv 1 \pmod{p^k}$ $\qquad\qquad(*)$

and we prove $\quad r^{(p-1)p^{k-1}} \not\equiv 1 \pmod{p^{k+1}}$.

Now g.c.d. (r, p^{k-1}) = g.c.d. $(r, p^k) = 1$, so by theorem 25
$$r^{(p-1)p^{k-2}} = r^{\varphi(p^{k-1})} \equiv 1 \pmod{p^{k-1}}$$
Hence there exists some integer a such that
$$r^{(p-1)p^{k-2}} = 1 + a \cdot p^{k-1}$$
and $p \nmid a$ since otherwise we would contradict our inductive assumption. Now raise both sides of this equation to the pth power and expand by the Binomial Theorem to get
$$[r^{(p-1)p^{k-2}}]^p = (1 + a \cdot p^{k-1})^p$$
$$r^{(p-1)p^{k-1}} = 1 + p \cdot a \cdot p^{k-1} + \tfrac{1}{2} \cdot p(p-1)a^2 \cdot p^{2k-2} + \ldots + a^p \cdot p^{pk-p}$$
$$\equiv 1 + a \cdot p^k \pmod{p^{k+1}}$$
since $k > 2$ all the terms after the first two are congruent to 0 mod p^{k+1}. Because $p \nmid a$, we see that $r^{(p-1)p^{k-1}} \not\equiv 1 \pmod{p^{k+1}}$ as we set out to prove. Hence by induction the equation is true for all k.

Now we show that this r is a primitive root of p^k. Let n be the order of r in U_{p^k}. Then n must divide $\varphi(p^k) = (p-1) \cdot p^{k-1}$. If $r^n \equiv 1 \pmod{p^k}$, then clearly $r^n \equiv 1 \pmod{p}$. Therefore we know that $(p-1) | n$ because r is a primitive root of p and so has order $p - 1$. Consequently n has the form $n = (p-1) \cdot p^m$, where $0 \le m \le k-1$. If it is the case that $n \ne (p-1) \cdot p^{k-1}$, we know $n | (p-1) \cdot p^{k-2}$; but if this is true, then $r^{(p-1) \cdot p^{k-2}} \equiv 1 \pmod{p^k}$, which contradicts the inductive assumption $(*)$. Therefore $n = (p-1) \cdot p^{k-1}$ and r is a primitive root of p^k.

It is easier to show that $2p^k$ has a primitive root.

Theorem 31: Let p be an odd prime. Then $2p^k$ has a primitive root for all $k \ge 1$.

Proof: Let r be a primitive root for p^k and we may as well assume r is odd because if it is even then $r + p^k$ is odd and is still a primitive root of p^k (they are congruent mod p^k). Hence we know that g.c.d.$(r, 2p^k) = 1$. So $r \in U_{2p^k}$ and the order of r mod $2p^k$ must divide $\varphi(2p^k) = \varphi(p^k) = (p-1) \cdot p^{k-1}$. However $r^n \equiv 1 \pmod{2p^k}$ implies that $r^n \equiv 1 \pmod{p^k}$. Therefore $\varphi(p^k) | n$ since r was chosen as a primitive root of p^k. Hence $n = \varphi(2p^k)$ and r is thus a primitive root for $2p^k$.

So if we need a primitive root for $2p^k$ we can reduce to the case of finding a primitive root for p^k. To find a primitive root for p^k we can reduce to the case of finding a primitive root for p^2 which reduces to the case of finding a primitive root for p.

In summary, the only integers for which a primitive root exists are 2, 4, p^k, and $2p^k$, where p is an odd prime and $k \ge 1$. These are the only times U_n is cyclic. In a later chapter we will determine the group structure of U_n for any value of n.

Exercise 56: Show that any primitive root of p^k is also a primitive root of p.

Exercise 57: Find the primitive roots (4 of them) of 26, and the primitive roots (8 of them) of 25.

Exercise 58: Find the primitive roots of 41 and 82.

Exercise 59: Show that 3 is a primitive root of all integers of the forms 7^k and $2 \cdot 7^k$.

Exercise 60: Find a number that is a primitive root of any integer of the form 17^k.

Lesson 18: The Index

We will take one more step in doing algebra in cyclic groups before we move on to non-Abelian groups. When U_n is cyclic, which is, when n has a primitive root, we can solve a great many congruence equations. We know that when n has a primitive root that $U_n = Z_{\varphi(n)}$. In the cyclic groups Z_n under additions, any equations we write are quite simple, but in $Z_{\varphi(n)}$ under multiplication the equations are anything but simple. This is our first opportunity to do algebra in the more traditional sense of solving an equation for x.

Each primitive root of n is a generator for U_n. No U_n has a unique primitive root; there is always a choice of which primitive root to use, but generators are all created equal at this point. Let $g \in U_n$ be an arbitrary element and let r be an arbitrary primitive root of n. Since $U_n = <r>$ there is some smallest positive integer power k of r such that $r^k = g$. This leads to the next definition.

Definition 19: Let n be a positive integer with a primitive root r, let g be any element of U_n. The smallest power k of r which equals g is called the **index of g relative to r**, and we write $k = ind_r(g)$.

It is important to keep in mind that the index is only defined if g is an element of U_n, so we must always know that g.c.d.(g,n) = 1 in order for the index to be a useful tool. We immediately know the following facts about the index:

1. $1 \leq ind_r(g) \leq \varphi(n)$.
2. indices that are congruent mod $\varphi(n)$ are equal.

Note especially that last fact. We will be solving congruence equations in U_n, that is, these equations will be written mod n. But using multiplication on U_n we will be solving those mod n congruences in $Z_{\varphi(n)}$ which is mod $\varphi(n)$. This can be a source of confusion at first. Since the index of g is an exponent, it behaves much like the logarithm function in the algebra you are familiar with, and it satisfies many of the same identities, as we show in the next theorem.

Theorem 32: a) $ind_r(gh) \equiv ind_r(g) + ind_r(h) \pmod{\varphi(n)}$.

b) $ind_r(g^k) \equiv k \cdot ind_r(g) \pmod{\varphi(n)}$.

c) $ind_r(1) \equiv 0 \pmod{\varphi(n)}$

d) $ind_r(r) \equiv 1 \pmod{\varphi(n)}$

Proof: only parts a) and b) need proving, because parts c) and d) are immediate.

To prove part a) let $ind_r(g) = n$ and $ind_r(h) = m$. This means that $r^n = g$ and $r^m = h$. Clearly then $g \cdot h = r^n \cdot r^m = r^{n+m}$ where the n+m must be taken mod $\varphi(n)$. Hence $ind_r(gh) \equiv n + k \pmod{\varphi(n)}$ which is what we wished to prove.

To prove part b) we do the same basic thing. Let $ind_r(g) = n$. Then $r^n = g$. Therefore $(r^n)^k = r^{nk}$, where nk must be taken mod $\varphi(n)$. Then $ind_r(g^k) \equiv nk \pmod{\varphi(n)}$, which is what we were to prove.

The index can be used, theoretically, to solve many – though not all - equations in U_n when n has a primitive root. For example, we will now solve $4x^9 \equiv 7 \pmod{13}$. In solving such a congruence equation, we always begin by finding a primitive root of n, in this case 13, and obtain the indices for the elements of U_{13}. For odd primes, 2 is usually a primitive root, though not always. In this case though the powers of 2 are easy to calculate:

$2^2 \equiv 4 \pmod{13}$ $\quad\quad$ $2^3 \equiv 8 \pmod{13}$ $\quad\quad$ $2^4 \equiv 3 \pmod{13}$ $\quad\quad$ $2^5 \equiv 6 \pmod{13}$

$2^6 \equiv 12 \pmod{13}$ $\quad$ $2^7 \equiv 11 \pmod{13}$ $\quad$ $2^8 \equiv 9 \pmod{13}$ $\quad$ $2^9 \equiv 5 \pmod{13}$
$2^{10} \equiv 10 \pmod{13}$ $\quad$ $2^{11} \equiv 7 \pmod{13}$ $\quad$ $2^{12} \equiv 1 \pmod{13}$

The indices are the exponents of the generator 2 and you can see that these generators form a cyclic group of order $12 = \varphi(13)$. Thus $\text{ind }_2(9) = 8$. To solve the equation $4x^9 \equiv 7 \pmod{13}$, we can now take the index of both sides of the equation and this changes it into a mod 12 congruence:

$\text{ind }_2(4 \cdot x^9) \equiv \text{ind }_2(7) \pmod{12}$

$\text{ind }_2(4) + 9 \cdot \text{ind }_2(x) \equiv \text{ind }_2(7) \pmod{12}$ $\quad\quad$ (using theorem 32a and 32b)

$2 + 9 \cdot \text{ind }_2(x) \equiv 11 \pmod{12}$

$9 \cdot \text{ind }_2(x) \equiv 9 \pmod{12}$

$\text{ind }_2(x) \equiv 1 \pmod{4}$ $\quad\quad$ (by theorem 19)

Now going back to mod 12, we find

$\text{ind }_2(x) \equiv 1, 5, \text{ or } 9 \pmod{12}$.

Therefore, by the table of indices above,

$x \equiv 2, 6, \text{ or } 5 \pmod{13}$.

This can be checked by substituting each solution into the equation, but it does take some effort. Choosing a different primitive root will lead to the same answer, and to be reassuring we will solve the same problem using one of the other primitive roots. By now you understand the structure of $\mathbf{Z}_{12}$ well enough that you know right away what the other primitive roots are: they must be 2^5, 2^7, and 2^{11}, which are 6, 11, and 7. Let's redo this problem using 7 as our primitive root.

$7^2 \equiv 10 \pmod{13}$ $\quad$ $7^3 \equiv 5 \pmod{13}$ $\quad$ $7^4 \equiv 9 \pmod{13}$ $\quad$ $7^5 \equiv 11 \pmod{13}$
$7^6 \equiv 12 \pmod{13}$ $\quad$ $7^7 \equiv 6 \pmod{13}$ $\quad$ $7^8 \equiv 3 \pmod{13}$ $\quad$ $7^9 \equiv 8 \pmod{13}$
$7^{10} \equiv 4 \pmod{13}$ $\quad$ $7^{11} \equiv 2 \pmod{13}$ $\quad$ $7^{12} \equiv 1 \pmod{13}$

Now take ind _7 of both sides:

$\text{ind }_7(4 \cdot x^9) \equiv \text{ind }_7(7) \pmod{12}$

$\text{ind }_7(4) + 9 \cdot \text{ind }_7(x) \equiv \text{ind }_7(7) \pmod{12}$ $\quad\quad$ (using theorem 32a and 32b)

$10 + 9 \cdot \text{ind }_7(x) \equiv 1 \pmod{12}$

$9 \cdot \text{ind }_7(x) \equiv -9 \pmod{12}$

$\text{ind }_7(x) \equiv -1 \pmod{4} \equiv 3 \pmod{4}$ $\quad\quad$ (by theorem 19)

Hence in mod 12, $\text{ind }_7(x) \equiv 3, 7, \text{ or } 11 \pmod{12}$ and so $x \equiv 5, 6, \text{ or } 2 \pmod{13}$, which are the same solutions naturally generated in a different order.

Let's try another more complicated example: solve $21x^6 \equiv 11 \pmod{50}$. $\varphi(50) = 20$ and we will try 3 as a primitive root, since 2 is not an element of $\mathbf{U}_{50}$.

$3^2 \equiv 9 \pmod{50}$ $\quad$ $3^3 \equiv 27 \pmod{50}$ $\quad$ $3^4 \equiv 31 \pmod{50}$ $\quad$ $3^5 \equiv 43 \pmod{50}$
$3^6 \equiv 29 \pmod{50}$ $\quad$ $3^7 \equiv 37 \pmod{50}$ $\quad$ $3^8 \equiv 11 \pmod{50}$ $\quad$ $3^9 \equiv 33 \pmod{50}$
$3^{10} \equiv 49 \pmod{50}$ $\quad$ $3^{11} \equiv 47 \pmod{50}$ $\quad$ $3^{12} \equiv 41 \pmod{50}$ $\quad$ $3^{13} \equiv 23 \pmod{50}$
$3^{14} \equiv 19 \pmod{50}$ $\quad$ $3^{15} \equiv 7 \pmod{50}$ $\quad$ $3^{16} \equiv 21 \pmod{50}$ $\quad$ $3^{17} \equiv 13 \pmod{50}$
$3^{18} \equiv 39 \pmod{50}$ $\quad$ $3^{19} \equiv 17 \pmod{50}$ $\quad$ $3^{20} \equiv 1 \pmod{50}$

Then take **ind $_3$** of both sides:

$$\text{ind}_3(21x^6) \equiv \text{ind}_3(11) \pmod{\varphi(50)}$$

$$\text{ind}_3(21) + 6 \cdot \text{ind}_3(x) \equiv \text{ind}_3(11) \pmod{20}$$

$$16 + 6 \cdot \text{ind}_3(x) \equiv 8 \pmod{20}$$

$$6 \cdot \text{ind}_3(x) \equiv 12 \pmod{20}$$

$$\text{ind}_3(x) \equiv 2 \pmod{10}$$

Hence **ind $_3$**$(x) \equiv 2$ or $12 \pmod{20}$. Therefore $x \equiv 9$ or $41 \pmod{50}$. These are easily checked, though it may take some time.

Notice that we are solving these equations in $\mathbf{U_n}$, and that means all the coefficients in the equation must be in $\mathbf{U_n}$; that is, to solve $a \cdot x^k \equiv b \pmod{n}$, not only must n have a primitive root, but by theorem 20 g.c.d. (a, n) must divide b; if not, the equation can have no solution.

Exercise 61: Solve, if possible, $7 \cdot x^3 \equiv 3 \pmod{11}$, $3 \cdot x^4 \equiv 5 \pmod{11}$, and $7 \cdot x^5 \equiv 4 \pmod{11}$.

Exercise 62: Solve, if possible, $x^{13} \equiv 13 \pmod{17}$, $8 \cdot x^5 \equiv 10 \pmod{17}$, and $9 \cdot x^8 \equiv 7 \pmod{17}$.

Exercise 63: Solve $4 \cdot x^5 \equiv 7 \pmod{27}$, $10 \cdot x^8 \equiv 16 \pmod{27}$, and $16 \cdot x^4 \equiv 13 \pmod{27}$, if possible.

Most of the material presented in these last nine lessons was first discovered by Carl Friedrich Gauss when he was 21 years old, though it wasn't published until three years later in what is now considered a classic of mathematics, *Disquisitiones Arithmeticae*, and he made these discoveries without the benefit of the modern algebraic tools we have used here. Gauss was born in 1777 in Brunswick in what is now Germany. He was a child prodigy, correcting his fathers arithmetic mistakes when he was three years old. The duke of Brunswick heard of his abilities and arranged for his higher education. Gauss was undecided whether to pursue a study of languages or mathematics, but some of his early mathematical discoveries so enthused him that he decided for math. His contributions both to mathematics and physics are too numerous to list here. Gauss continued mentally active and mathematically brilliant until his death from a heart attack in 1855. He ranks, along with Euler, as one of the most brilliant mathematicians who have ever lived.

"When I have clarified and exhausted a subject, then I turn away from it in order to go into darkness again."

Lesson 19: Permutations and the Symmetric Groups

We will now pause in our study of Abelian groups and devote some attention to the non-Abelian groups. Much of the next six lessons are due to Cayley, as mentioned earlier. We made considerable progress understanding the family of cyclic groups, but progress comes more slowly with non-Abelian groups. We will make rather less progress but on several different families of groups. Though we can't yet delve into the details of these groups, we can introduce many of the tools we will need to proceed to those details. Since we can't use the commutative law we naturally expect these groups to be more difficult to study. Not only that, but the entire character of the subject matter changes radically. We will leave off solving of equations and do work that is very different from what we have been doing.

To get into non-Abelian groups we will go back to the only example of a non-Abelian group that we have met so far. In lesson 8, as an example, we looked at the set of all bijections from a set S onto itself and noted that this set of all bijections under the operation of composition formed a non-Abelian group. At this point, we will change our terminology. Also our notation:iIt is customary to use Greek letters as the names for permutations.

Definition 20: Let B be any set. A bijection from B onto itself, $\sigma: B \to B$, is called a **permutation of B**.

From lesson 8 we know that the set of all permutations of the set A under the operation of composition forms a group. Our focus here is on the permutations of A, not on the elements of A. We will usually not care what the elements of A are, but we will very much care about how many elements A has. If A has n elements, then we will refer to the group of all permutations of A as S_n. We have not yet defined what it means for two groups to be the same so you must take it on faith that changing the elements in the set has no effect on the nature of the group of permutations. For clarity we will reserve the word "element" exclusively for the permutations, the elements of the group S_n, and not use it for the elements of the set A. Instead we will refer to the elements of the set A as *letters*, even when they are actually numerals.

Definition 21: The group S_n of all the permutations on a set of n letters is called the **symmetric group on n letters**.

In lesson 8 we represented the bijections as function diagrams. That is an awkward way of writing the elements of these groups and we will want a better one. To get to a better notation, let's

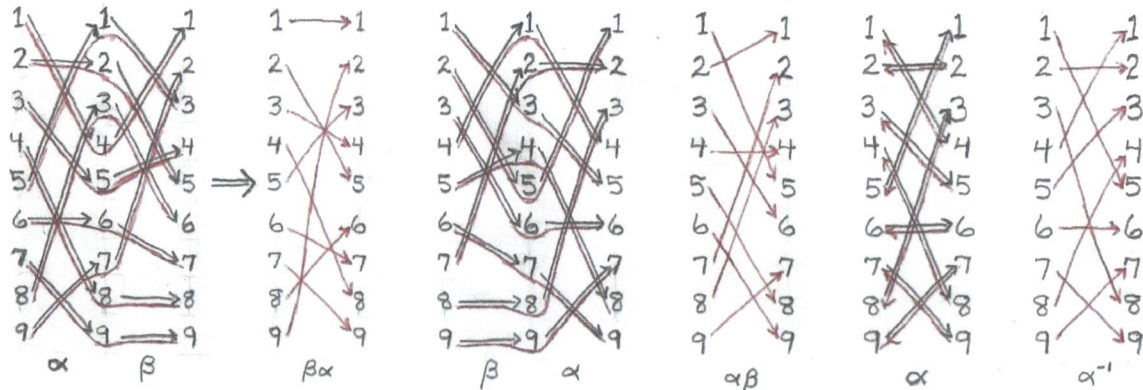

review how we used function diagrams. Consider the following permutations defined in the diagrams on the preceding page. We compute the compositions of functions simply by following arrows. The only trick to remember is that the diagrams are written in the reverse order of the way compositions are written. Thus, βα is the composition created by first applying α and then applying β, the opposite of the way they are pictured in the function diagrams. It is obvious from the diagrams that βα ≠ αβ, so we know **S$_n$** is indeed a non-Abelian group.

Note that we can compute the inverse of α by simply turning all the arrows to point in the opposite direction. Of course then we would also want to flip the left and right sides so that the arrows still point from left to right. Study the function diagrams until you fully understand how to use them.

Function diagrams are useful enough when n is small, but even when n is small they are rather awkward. We really need to devise a more compact way of writing permutations, but a new way that is just as easy to read as a function diagram. There are, in fact, two other ways of writing permutations that are much more convenient. In this lesson we will learn what is called *permutation notation*.

The basic idea of permutation notation is to make the diagram horizontal rather than vertical, and to eliminate the arrows. Rather than arrows to show where an element is mapped, we simply write its image underneath it. Then we enclose the whole correspondence in parentheses. Here is how α, and β look in permutation notation. Compare the function diagram given above with the permutation notation and be sure you understand how the two are the same and how to read the permutation notation.

Unfortunately it is a bit trickier to multiply two permutations written this way. For one thing, now they are

$$\alpha = \begin{pmatrix} 1 & 2 & 3 & 4 & 5 & 6 & 7 & 8 & 9 \\ 4 & 2 & 5 & 8 & 1 & 6 & 9 & 3 & 7 \end{pmatrix} \quad \beta = \begin{pmatrix} 1 & 2 & 3 & 4 & 5 & 6 & 7 & 8 & 9 \\ 3 & 5 & 6 & 1 & 4 & 7 & 2 & 8 & 9 \end{pmatrix}$$

horizontal, we write them in the usual algebraic order, so that when we do multiply them we must remember to start with the one on the right. Otherwise, multiplication of permutations in this notation is much like following arrows. To compute βα let's write the two permutations side and side, with β on the left, and follow how it is done. I have drawn in arrows for a few of the letters to show how the process goes.

We can also easily compute the inverse of a permutation using a two step procedure.

$$\beta\alpha = \begin{pmatrix} 1 & 2 & 3 & 4 & 5 & 6 & 7 & 8 & 9 \\ 3 & 5 & 6 & 1 & 4 & 7 & 2 & 8 & 9 \end{pmatrix}\begin{pmatrix} 1 & 2 & 3 & 4 & 5 & 6 & 7 & 8 & 9 \\ 4 & 2 & 5 & 8 & 1 & 6 & 9 & 3 & 7 \end{pmatrix}$$

$$= \begin{pmatrix} 1 & 2 & 3 & 4 & 5 & 6 & 7 & 8 & 9 \\ 1 & 5 & 4 & 8 & 3 & 7 & 9 & 6 & 2 \end{pmatrix}$$

Step 1: simply turn the permutation upside down. The top row is now what was the bottom row, but the entries are no longer in the correct numerical order.

Step 2: rearrange the columns so that the entries in the top row will be in the correct numerical order. The diagram below shows how α^{-1} is computed with these steps.

$$\alpha = \begin{pmatrix} 1 & 2 & 3 & 4 & 5 & 6 & 7 & 8 & 9 \\ 4 & 2 & 5 & 8 & 1 & 6 & 9 & 3 & 7 \end{pmatrix} \rightarrow \alpha^{-1} = \begin{pmatrix} 4 & 2 & 5 & 8 & 1 & 6 & 9 & 3 & 7 \\ 1 & 2 & 3 & 4 & 5 & 6 & 7 & 8 & 9 \end{pmatrix} = \begin{pmatrix} 1 & 2 & 3 & 4 & 5 & 6 & 7 & 8 & 9 \\ 5 & 2 & 8 & 1 & 3 & 6 & 9 & 4 & 7 \end{pmatrix}$$

When a permutation maps one letter to any other letter, we say that the permutation *moves* the letter. Thus the permutation α moves the letters 1, 3, 4, 5, 7, 8, and 9. When the permutation maps a letter to itself, we say that the permutation *fixes* the letter.

Now that we can do some computations, it is time to step back and consider what exactly we have. For a given set of n letters, A, we have a non-Abelian group, **S$_n$**. Just how big a group is **S$_n$**? We can easily count the number of possible permutations of n letters. First write down 1 through n in order. Now how many letters can we put under the 1? Obviously any of the n. Once we have made that choice, how many letters can we put under the 2? Any of the n – 1 letters that are left unused. So there are n(n – 1) possible ways to make these first two choices. Continuing in this way, there are n – 2 possible letters to go under the 3, n – 3 possible letters to go under the 4, and so on until there is only 1 possible letter to go under the n. So we see that there are n(n – 1)(n – 2)...4·3·2·1 = n!. These groups quickly become enormous. The order of S_3 is 3! = 6, a second group besides Z_6 of order 6 and this time non-Abelian. The order of S_4 is 4! = 24, the order of S_5 is 60, and we are already beyond what we can handle reasonably easily. With Abelian groups we started with small groups and got larger; with non-Abelian groups we start with mainly gigantic groups and hope to find some smaller ones.

Exercise 64: List all the elements of S_3 and of S_4 as both function diagrams and in permutation notation.

Exercise 65: For the permutations listed below, compute the following: $\zeta\eta$, $\eta\theta$, $\zeta\theta$, $\zeta\eta\theta$, ζ^2, η^{-1}, θ^{-2}, $\eta^{-1}\zeta\eta$. Note that these permutations are all elements of S_{10} since they all involve 10 letters.

Exercise 66: Find $<\zeta>$, $<\eta>$, $<\theta>$, $<\zeta>\cap<\eta>$, $<\theta>\cap<\eta>$, and $<\zeta>\cap<\theta>$.

$$\zeta = \begin{pmatrix} 0 & 1 & 2 & 3 & 4 & 5 & 6 & 7 & 8 & 9 \\ 3 & 4 & 9 & 2 & 0 & 5 & 1 & 7 & 6 & 8 \end{pmatrix} \quad \eta = \begin{pmatrix} 0 & 1 & 2 & 3 & 4 & 5 & 6 & 7 & 8 & 9 \\ 1 & 2 & 3 & 4 & 5 & 6 & 7 & 8 & 9 & 0 \end{pmatrix} \quad \theta = \begin{pmatrix} 0 & 1 & 2 & 3 & 4 & 5 & 6 & 7 & 8 & 9 \\ 4 & 6 & 0 & 1 & 5 & 7 & 3 & 2 & 9 & 8 \end{pmatrix}$$

Lesson 20: Cycles

The permutation notation we introduced in lesson 19 is a big improvement over function diagrams, but we can do better. Finding a suitable way of writing our ideas is critical for making further progress in understanding. The notation we will learn in this lesson is called *cycle notation*. It has one disadvantage over permutation notation, but its advantages far outweigh the disadvantage.

The idea of cycle notation is to follow one letter at a time and see how it is moved around by repeated application of the permutation. Since there are only finitely many letters, at this stage of our study, it is clear that at some point the letter must return to itself. More applications of the permutation will make that letter repeat its journey through the other letters and we will just keep going around. This is similar to the process of computing the cyclic subgroup generated by an element, and so it is natural to use similar terminology. We will begin with these three permutations:

$$\alpha = \begin{pmatrix} 1 & 2 & 3 & 4 & 5 & 6 & 7 & 8 & 9 \\ 4 & 2 & 5 & 8 & 1 & 6 & 9 & 3 & 7 \end{pmatrix} \quad \beta = \begin{pmatrix} 1 & 2 & 3 & 4 & 5 & 6 & 7 & 8 & 9 \\ 3 & 5 & 6 & 1 & 4 & 7 & 2 & 8 & 9 \end{pmatrix} \quad \gamma = \begin{pmatrix} 1 & 2 & 3 & 4 & 5 & 6 & 7 & 8 & 9 \\ 1 & 3 & 2 & 5 & 6 & 4 & 9 & 8 & 7 \end{pmatrix}$$

It is most natural to begin by following the letter 1. In α, 1 is moved to 4. A second application of α moves 4 to 8; so far we have 1 → 4 → 8. A third application of α moves 8 to 3; another application of α then moves 3 to 5; then 5 is moved back to where we started, to 1. Clearly more applications of α would repeat this sequence of letters. As α is applied repeatedly to 1, we get 1 → 4 → 8 → 3 → 5 → 1. Rather than using arrows, we will enclose the whole thing in parentheses and put in commas to separate the letters. So the effect of α on the letter 1 is written (1, 4, 8, 3, 5). This is called a *cycle*. We close the parentheses without repeating the letter we started with; it is unnecessary to repeat the 1 as long as we remember that closing the parenthesis means to start over. But this does not represent all that α does. What about the letter 2? α fixes 2 and we can indicate that by (2). The next letter that we haven't yet followed is 6, also fixed. But 7 is not fixed and has not yet been followed. You should easily see that α moves 7 to 9 and then moves 9 back to 7, so we will write this as (7, 9). Now we have accounted for all the letters so we write these cycles all together. It is customary to put the longer cycles to the left, so α = (1, 4, 8, 3, 5)(7, 9)(2)(8). This is the cycle notation for α.

In the same way be sure you can come up with the cycle notation for β and γ, namely

β = (1, 3, 6, 7, 2, 5, 4)(8)(9)   and   γ = (4, 5, 6)(2, 3)(7, 9)(1)(8)

This cycle notation for writing a permutation tells us everything that the permutation notation told us; namely, where does a permutation move a certain letter? Answer: to the letter immediately to the right of it in its cycle, or else to the first letter of the cycle if it is the last letter in the cycle. That is really all the information we asked of the permutation notation. The cycle notation is shorter, and by a glad coincidence it gives us even more information than the permutation notation did. The cycles reveal the internal working of the permutation. For example, the cycle notation for β tells us that 8 and 9 are fixed but the other seven letters are all part of a single cycle. γ on the other hand fixes 1 and 8, then there are three letters that are moved around in a triangular way, and two pairs of elements that are just interchanged with each other. This is all information that is accessible from the permutation notation, but there you have to look for it, and in the cycle notation it is plain to see.

We can draw a picture of what the cycle notation tells us. On the next page are diagrams that show α, β, γ and also α^{-1}. The cycle notation is thus much more revealing than the permutation notation was. There are, however, some drawbacks to the notation. The most significant problem is that cycle notation is not unique as the permutation notation is. For any particular cycle, we can write it in many ways that may seem, at first glance, to mean different cycles. The problem is that "a circle has no beginning", as they say. The cycle in α, (1, 4, 8, 3, 5), might just as well begin at 4, in which case it would be written (4, 8, 3, 5, 1), or with 8, in which case it is (8, 3, 5, 1, 4). The notation requires that we notice for ourselves when two cycles are really the same. It is only the order of the letters that matters, not where it begins.

It is common practice to not write down elements that are fixed by a permutation when we are using the cycle notation; unless not writing them down will cause confusion. For example, in permutation notation we know that α, β, and γ are all understood as being elements of S_9. But when we write α in cycle notation, suppressing the fixed letters, we have α = (1, 4, 8, 3, 5)(7, 9) and only seven letters appear. Hence we can think of α as an element of S_7, but we will have to indicate the missing letters if we want to view it as an element of S_9.

It is very easy to find the inverse of a cycle; just write the same letters in the opposite order. It doesn't matter which letter you begin with, just change the order from left to right into right to left. If a permutation is composed of several cycles, reverse each one. It should be obvious that if a cycle has length 2, then it is its own inverse. Thus we find

α = (1, 4, 8, 3, 5)(7, 9)(2)(8) ⟹ $α^{-1}$ = (1, 5, 3, 8, 4)(7, 9)(2)(8)

β = (1, 3, 6, 7, 2, 5, 4)(8)(9) ⟹ $β^{-1}$ = (4, 5, 2, 7, 6, 3, 1)(8)(9)

γ = (4, 5, 6)(2, 3)(7, 9)(1)(8) ⟹ $γ^{-1}$ = (4, 6, 5)(2, 3)(9, 7)(1)(8)

Different cycles involve different numbers of letters. The number of letters in a particular cycle is referred to as its *length*. Thus we can say that α is composed of a cycle of length 5, a cycle of length 2, and two cycles of length 1.

It is important to notice this fact: when we are dealing with a single permutation, one that we can write in permutation notation with a single set of parentheses, then when it is changed to cycle notation the cycles will be *disjoint* from each other; there will be no letters shared between the different cycles. Disjoint cycles are entirely independent of each other; they have no effect on each other as you can see immediately in the diagrams given above. A is composed of two disjoint cycles and γ is composed of three disjoint cycles. Since the cycles are entirely independent of one another, it makes no difference which order they are written in; that is, they commute with one another. Thus

γ = (4, 5, 6)(2, 3)(7, 9) = (2, 3)(4, 5, 6)(7, 9) = (7, 9)(2, 3)(4, 5, 6) = (7, 9)(4, 5, 6)(2, 3)

Anytime there we have disjoint cycles, they can be multiplied in any order, but it is usual to write the longer cycles to the left.

It is only when the cycles are not disjoint that the order of writing them is critical. Whenever there are cycles that share one or more letters, those cycles come from different permutations and should be multiplied to determine which particular permutation the cycles represent. We will discuss how to multiply cycles in the next lesson.

We must now introduce more technical terms.

Definition 22: A cycle of length n is called an **n-cycle**.

Definition 23: A 2-cycle is called a **transposition**.

In the previous lesson, you were asked to compute the subgroup generated by various permutations. With cycle notation, it is even easier to find the subgroup, with the additional advantage that you know ahead of time how many elements the subgroup will contain. It is should be ckear that an n-cycle will generate a subgroup of order n. Or, in other words, an n-cycle generates a copy of Z_n. Or, in still other words, an n-cycle is an element of order n.

But what of permutations that are composed of more than one cycle? Only β is a single 7 cycle, so we know that $<\beta> = \mathbf{Z}_7$. What is the order of α or γ? The answer is given by a theorem whose proof will be left as an exercise.

Theorem 33: The order of a permutation, σ, equals the least common multiple of the lengths of its cycles.

Proof: is an exercise.

Thus we can immediately see that the order of α is the least common multiple of 5 and 2, which is 10, and the order of γ is the least common multiple of 3, 2, and 2, which is 6.

Exercise 67: Write all the elements of S_3 and S_4 in cycle notation.

Exercise 68: Prove theorem 33.

Exercise 69: Change all the permutations in exercise 65 into cycle notation.

Lesson 21: Some Cycle Identities

We will now develop some skill at multiplying cycles together. Multiplication of cycles is only defined or necessary when the cycles are not disjoint. Usually we will want to multiply the overlapping cycles to produce a permutation written using only disjoint cycles. However there are times when it is important to do the opposite, to factor a given permutation into a product of overlapping cycles. There are two forms in particular that we will need to factor permutations into in the future. We will also find it handy to be able to rearrange a factorization to make a particular cycle come first.

To multiply a series of non-disjoint cycles, choose a letter to follow, say x_1. Begin with the cycle on the right and find the letter it maps x_1 to; say x_2. Take x_2 into the next cycle to the left and find where it maps x_2; if x_2 does not occur in that cycle, go on to the next cycle to the left, and keep going until you find a cycle that does include x_2; find where x_2 is mapped by that cycle, say x_3. Take x_3 and move on to the next cycle to the left that includes it, and find where it goes, say x_4. When you reach the last of the cycles, write down the letter you began with, a comma, and the letter you reached in the leftmost cycle, say x_n: $(x_1, x_n$. Now repeat the process beginning with x_n and with the cycle furthest to the right; follow through all the changes x_n goes through just as you did with x_1 until you again reach the end, this time resulting in x_m: now write $(x_1, x_n, x_m$. Repeat the process until the cycle closes. Begin a new cycle with some letter not yet included in your result. Once you have followed every letter, you are done. As an example, be sure you can see how these multiplications are accomplished:

$(2, 7, 9, 1, 3, 6)(1, 7, 4, 9) = (1, 9, 3, 6, 2, 7, 4)$

$(2, 4, 7)(3, 7)(4, 6, 2, 1)(2, 9) = (1, 7, 3, 2, 9)(4, 6)$

It is as important to be able to factor a cycle as it is to multiply them. First let's consider how to write a cycle of any length into the product of transpositions. It should be obvious that any permutations can be written that way. A permutation is essentially just a rearrangement of the letters, and we can get any rearrangement we want by switching two letters at a time. Intuitively this is obvious. We will prove this as a theorem:

Theorem 34: $(a_1, a_2, a_3, \ldots, a_{n-1}, a_n) = (a_1, a_n)(a_1, a_{n-1})\ldots(a_1, a_3)(a_1, a_2)$.

Proof: The proof is a straightforward matter of multiplying the transpositions on the right hand side of the equation and observing that the result is equal to the left hand side.

For example, let's rewrite a cycle in terms of transpositions using theorem 34, and illustrate the inner workings with a function diagram:

$\sigma = (1, 6, 5, 7, 4, 2, 3) = (1, 3)(1, 2)(1, 4)(1, 7)(1, 5)(1, 6)$

Note that we could have done this in many different ways by starting the original cycle at a different letter:

$\sigma = (1, 6, 5, 7, 4, 2, 3) = (7, 4, 2, 3, 1, 6, 5) = (7, 5)(7, 6)(7, 1)(7, 3)(7, 2)(7, 4)$

Thus we have two completely distinct ways of factoring σ but both of them are valid. The only way to tell that the two factorizations represent the same permutation is by multiplying them out and seeing that they are the same. This is why we will typically insist that a permutation should always be written in "lowest terms", as a product of disjoint cycles.

It is also worth noting that an n-cycle will be expressed as the product of $n - 1$ transpositions when we use theorem 34 to do it. This is a useful fact to know because the number of transpositions in a factorization is sometimes more important to know than what the actual transpositions are.

Now while we can split up any permutation into a product of transpositions in many different ways, there are some things that all such factorizations have in common. For a given permutation, the factorizations always involve an even number of transpositions, or they always involve an odd number of transpositions.

Theorem 35: Let σ be any permutation that can be factored into an even number of transpositions. Then every factorization of σ yields an even number of transpositions.

Proof: Here I will give one of the traditional proofs. It is not straight-forward, but it is a typical example of one important way of to think about permutations.

Suppose we are working with n letters. Whatever the letters are, replace them with n random positive integers and label them as $x_1, x_2, x_3, \ldots, x_{n-1}, x_n$. We will not be concerned with what these numbers are; we will only be concerned with the subscripts of x that are associated with them. Any permutation of the original letters will permute these numbers by permuting their subscripts. Now from these numbers we will compute a large product, call it D; each factor in the product is chosen to be one of the numbers minus another of the numbers with a larger subscript. Thus, the factors will include $(x_1 - x_3)$ but not $(x_3 - x_1)$ and not $(x_3 - x_3)$ – which, of course, is 0. Since we have chosen the n numbers in an arbitrary fashion, some of these factors may be negative and some may be positive. What we are interested in is this: if we let a random transposition, say (x_i, x_j), permute the subscripts, how will that change the product D?

To see clearly what happens, here is the complete product, D, written out in an orderly fashion, with the subscript of the first term in each factor determining the row, and the subscript of the second term in each factor determining the column. We may assume, without loss of generality, that $i < j$. The reason for the coloring of various factors will be explained shortly.

$$D = (x_1-x_2)(x_1-x_3)\cdots(x_1-x_{i-1})(x_1-x_i)(x_1-x_{i+1})\cdots(x_1-x_{j-1})(x_1-x_j)(x_1-x_{j+1})\cdots(x_1-x_{n-1})(x_1-x_n)\cdot$$
$$(x_2-x_3)\cdots(x_2-x_{i-1})(x_2-x_i)(x_2-x_{i+1})\cdots(x_2-x_{j-1})(x_2-x_j)(x_2-x_{j+1})\cdots(x_2-x_{n-1})(x_2-x_n)\cdot$$
$$\cdots$$
$$(x_{i-1}-x_i)(x_{i-1}-x_{i+1})\cdots(x_{i-1}-x_{j-1})(x_{i-1}-x_j)(x_{i-1}-x_{j+1})\cdots(x_{i-1}-x_{n-1})(x_{i-1}-x_n)\cdot$$
$$(x_i-x_{i+1})\cdots(x_i-x_{j-1})(x_i-x_j)(x_i-x_{j+1})\cdots(x_i-x_{n-1})(x_i-x_n)\cdot$$
$$(x_{i+1}-x_{i+2})\cdots(x_{i+1}-x_{j-1})(x_{i+1}-x_j)(x_{i+1}-x_{j+1})\cdots(x_{i+1}-x_{n-1})(x_{i+1}-x_n)\cdot$$
$$\cdots$$
$$(x_{j-1}-x_j)(x_{j-1}-x_{j+1})\cdots(x_{j-1}-x_{n-1})(x_{j-1}-x_n)\cdot$$
$$(x_j-x_{j+1})\cdots(x_j-x_{n-1})(x_j-x_n)\cdot$$
$$\cdots$$
$$(x_{n-2}-x_{n-1})(x_{n-2}-x_n)\cdot$$
$$(x_{n-1}-x_n)$$

In considering how the transposition (i, j) alters D, note that it simply interchanges x_i with x_j and nothing more. In particular, (i, j) does not change x_{i+1} to x_{j+1}, since they are not the letters being permuted.

Very well then, what is the result of applying (i,j) to D, which we will write $(i,j)D$? An inspection of the various factors in the product above, shows that the factors highlighted in green are simply interchanged with each other, and have no net effect on D. Similarly, the factors highlighted in blue are interchanged with each other and result in no net change in D. But the factors highlighted in yellow are interchanged with each other and each of these interchanges changes the sign of D twice; hence there is no net effect on the value of D. The unhighlighted factors are left alone. Only the factor highlighted in pink, $(x_i - x_j)$, is not interchanged with some other factor. It is changed to

(x_j–x_i), which is the negative of what it was. Therefore the cumulative effect of a transposition acting on D is to change it into –D.

Therefore if σ can be factored into an even number of transpositions, the net effect of letting σ act on D would be to leave D entirely unaltered; but if σ can be factored into an odd number of transpositions, the net effect would be to change D into –D. Therefore, if σ can be factored into an even number of transpositions, then it cannot be factored into an odd number of transpositions, and vice versa.

This theorem just begs to be formalized in a definition.

Definition 24: If a permutation σ can be factored into an even number of transpositions then it is called an **even permutation**, and if σ can be factored into an odd number of transpositions then it is called an **odd permutation**.

There is an important fact that is an immediate result of the preceding theorem which I will give here as a corollary.

Corollary 35a: If σ is an even permutation, then it can be factored into a product of 3-cycles only. If σ is an odd permutation, it can be factored completely into 3-cycles except for a single transposition.

Proof: Simply note the following two permutation identities:
$$(a, b)(a, c) = (a, c, b) \text{ and } (a, b)(c, d) = (a, d, c)(a, b, c)$$

These are easily verified. An even permutation can be factored into some number of pairs of transpositions. When the even number of permutations are associated into adjoining pairs, the pairs will either be disjoint or not. In either case, the above identities change them into 3-cycles. An odd permutation can be factored into some number of pairs of transpositions plus one more transposition. Rewrite the pairs as 3-cycles, and there will be one transposition remaining.

None of these factorizations will be unique. It is worth remembering that an n-cycle is an even permutation when n is odd, and it is an odd permutation when n is even. When a permutation is a product of two or more disjoint cycles, factor the permutation by factoring each of its cycles separately.

Exercise 70: Factor all the permutations in exercise 65 into transpositions in two different ways, and use each factorization to factor them into 3-cycles in two different ways, with a possible single transposition left over.

Exercise 71: Factor the following two permutations into transpositions in two different ways, and then factor them into 3-cycles and at most one transposition in two different ways.
$$(1, 7, 6, 4, 2)(3, 8, 11, 12, 5) \text{ and } (1, 10, 4, 6, 2, 7)(9, 3, 11, 5, 12, 8)$$

Lesson 22: The Alternating Groups

In order to understand the symmetric groups, S_n, we will want to make some progress in finding their lattice diagrams. Unfortunately, the symmetric groups quickly become much too large to make this a realistic goal. By the time n = 5, the lattice diagram for S_5 has become too formidable for all but the most determined. We can, however, get a great deal of information from a partial lattice diagram, and thanks to theorem 35 we have a beginning. It is a short step from theorem 35 to the following result.

Theorem 36: The set of all even permutations of n letters forms a group. As a subgroup of S_n, it includes exactly half of the permutations of S_n, so that its order is ½·n!.

Proof: We may assume n ≥ 4; for n = 3, we can simply list all the premutations and note that the theorem is true for this case. Let E be the set of all even permutations on n letters. So we can write E = $\{\sigma_1, \sigma_2, \sigma_3, \sigma_4, \ldots, \sigma_k\}$. Now choose any transposition, say (1, 3) and multiply each element of E by (1, 3) on the left. The result is k distinct odd permutations for if we had (1, 3)σ_i = (1,3)σ_j then multiplying again by (1, 3) on the left would yield $\sigma_i = \sigma_j$, a contradiction. Therefore there are at least as many odd permutations as there are even permutations. We can repeat the same argument beginning with the set of all odd permutations to show that there are at least as many even permutations as there are odd permutations. Therefore exactly half of the permutations on n letters are even.

Definition 25: The group of even permutations on n letters is called **the alternating group on n letters** and is denoted by A_n.

Thus we know that every symmetric group has a subgroup half its order which is the alternating group on those letters. This is not much of a start on finding the lattice diagram, but it is a start. There is one symmetric group that is small enough to diagram easily, S_3, since it has only 3! = 6 elements. We can list them all with no trouble:

(1)   (1, 2)   (1, 3)   (2, 3)   (1, 2, 3)   (1, 3, 2)

The cyclic subgroups are easy to find. The transpositions clearly generate subgroups of order 2 since they have order 2. This gives us three copies of Z_2. A 3-cycle will generate a subgroup of order 3:

< (1, 2, 3) > = { (1, 2, 3), (1, 2, 3)^2 = (1, 3, 2), (1, 2, 3)^3 = (1) }.

Both of the 3-cycles are in the same cyclic subgroup. It is easy to check that the three transpositions do not form a Klein 4-group, so it appears we have found all of the subgroups of S_3. There is one thing that is important to notice, though. The subgroup of even permutations is just the identity and the 3-cycles. Thus A_3 is the same group as Z_3. We still haven't defined what it means for two groups to be the same, but it should be clear enough what we will mean when we do define it. Two groups are the same if the only difference between them is the names we give to their elements.

We will now approach diagramming S_4 by first producing the lattice diagram of A_4. S_4 has order 24 and even this early in the family is quite formidable. A_4 is only order 12 and can be done easily with a little care. First, we must list the even permutations on four letters. 4-cycles are all odd, so the even permutations are either 3-cycles (all even) or products of two disjoint transpositions. Once we realize this is it easy to list all the even permutations:

(1, 2, 3)   (1, 3, 2)   (1, 2, 4)   (1, 4, 2)   (1, 3, 4)   (1, 4, 3)   (2, 3, 4)   (2, 4, 3)
(1, 2)(3, 4)   (1, 3)(2, 4)   (1, 4)(2, 3)

Including the identity, this is all of the expected twelve permutations. Now a 3-cycle is order 3, so we know each one is going to generate a copy of Z_3. Each copy of Z_3 includes two 3-cycles. Thus we see that we have found four copies of Z_3 and it is easy to list out the elements of each one:

first copy of Z_3 is   { (1), (1, 2, 3), (1, 3, 2) }
second copy of Z_3 is   { (1), (1, 2, 4), (1, 4, 2) }

third copy of Z_3 is { (1), (1, 3, 4), (1, 4, 3) }
fourth copy of Z_3 is { (1), (2, 3, 4), (2, 4, 3) }

The three transpositions each have order 2 and so they make three distinct copies of Z_2. The only two things left to check is whether they collectively form a copy of V, and whether we can find any subgroup of order 6. To check for V just multiply any two elements and see if the answer is the third element:

$$(1, 2)(3, 4)\ (1, 3)(2, 4) = (1, 4)(2, 3)$$

so these three order 2 elements do make a copy of V.

There is obviously no subgroup that is the same as Z_6 because that would require a 6-cycle. We are not done looking for subgroups of order 6, however. We now know that S_3 is a group of order 6 so we must ask if there might be a copy of the symmetric group on three letters present. Looking at the structure of S_3 we see that it needs a copy of Z_3 and three copies of Z_2. That seems possible until we notice that the three elements of order 2 in A_4 form a copy of V, but the three elements of order 2 in S_3 do not. Therefore there is no subgroup that looks like S_3 in A_4. It is possible there are other subgroups of A_4 that we do not know about yet and hence can't look for. Thus we have done all we can and the best lattice diagram we can produce for A_4 is shown above.

To diagram S_4 requires a great deal more work even though we only need to consider the odd permutations to complete it. S_4 has several subgroups we haven't learned about yet, and it has subgroups that include both even and odd permutations.

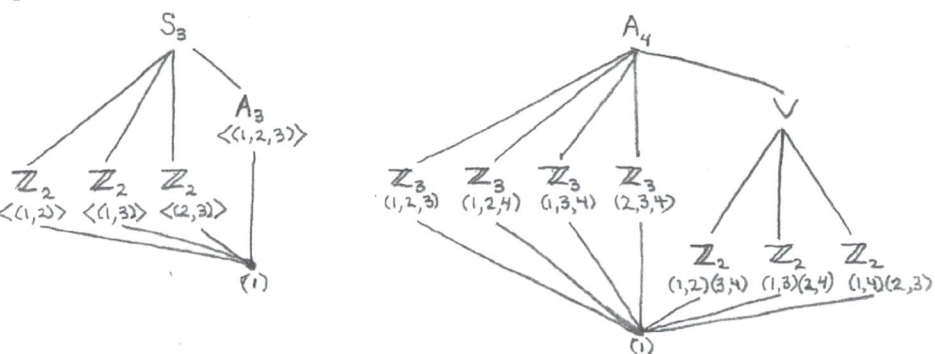

What we do know is that the entire lattice diagram of A_4 as shown above is included in the lattice diagram of S_4. We can still find the cyclic subgroups that are in S_4 that are generated by the odd permutations and insert those along with everything in A_4. In the next lesson we will discuss how to recognize other easy-to-spot subgroups of large symmetric groups.

Exercise 72: Find all of the remaining cyclic subgroups of S_4 and arrange them in a more complete lattice diagram. Include all of A_4 as well, of course.

Exercise 73: List all the elements of A_5; that is, list all the even permutations of five letters. (There will be sixty of them, including the identity permutation.) Make a stab at a lattice diagram for A_5; include at least the cyclic subgroups.

Lesson 23: Stabilizers

Though symmetric groups on n letters become horrifically large for fairly small values of n, there are some significant subgroups that are easy to find. To see why some subgroups are so easy to find, consider what a permutation of n letters may look like. Every permutation will move some letters and fix other letters. Let's focus for the moment on a single letter, say the letter 1. Each permutation in S_n will either move 1 or it will fix 1. Permutations that move the letter 1 can move it in n–1 different ways depending on which other letter it moves 1 to. But there is only one way to fix a letter. This suggests the idea of collecting together all the permutations of S_n that fix a particular letter.

Definition 26: Let k be a letter permuted by S_n. The set $Stab(k) = \{\sigma \in S_n: \sigma \text{ fixes } k\}$ is called the **stabilizer of k**. More generally, let X be a set of all the letters moved by some element of S_n. Then $Stab(X) = \{\sigma \in S_n: \sigma(x) = x \ \forall x \in X\}$ is called the **stabilizer of X**.

We are interested in stabilizers primarily because of the next theorem.

Theorem 37: Let X be any subset of letters permuted by S_n. Then $Stab(X) \leq S_n$.

Proof: Clearly the identity permutation is an element of the stabilizer of X. If σ is in the stabilizer of X, then σ^{-1} must also fix every letter in X. If τ is any other permutation that fixes every letter of X, then clearly στ will also fix every letter of X. Therefore $Stab(X)$ is a subgroup of S_n.

....

To begin the discussion, let's concentrate on the case where X consists in a single letter. Let the letters that are permuted by S_n be {1, 2, 3, …, n}. We can equally well choose any letter to focus on so we will consider $Stab(1)$. The only restriction on a permutation being an element of $Stab(1)$ is that it fixes 1; it is free to do anything at all with the other letters. The collection of all the permutations that fix 1 but move the other n–1 letters in an arbitrary manner is therefore identical to the group S_{n-1}. Similarly, $Stab(2)$ will be identical to S_{n-1}. $Stab(1)$ and $Stab(2)$ are identical as groups, but they are not identical to each other as subgroups. Since they fix different letters, they include different permutations from S_n.

Hence we see that S_n must contain n copies of subgroups that are identical to S_{n-1}. It follows that each subgroup copy of S_{n-1} must have n – 1 subgroups that are all the same as S_{n-2}. We might at first suppose that this will give us n·(n – 1) copies of S_{n-2} in S_n but we must think about it more carefully. Might there not be some copies of S_{n-2} that are contained in more that one copy of S_{n-1}? Of course there are. The permutations of a copy of S_{n-2} will be those permutations which fix two particular letters, say 1 and 2. Then clearly this S_{n-2} will be a subgroup of the copy of S_{n-1} which fixes 1 and also a subgroup of the copy of S_{n-1} which fixes 2. Clearly, each copy of S_{n-2} will be contained in two different copies of S_{n-1} and so there will be only ½·n·(n–1) copies of S_{n-2} in S_n.

If you have studied combinatorics then you will recognize this formula as the number of ways of choosing two objects out of n total objects, denoted by the combination symbol, $_nC_2$, and once you recognize this fact it makes complete sense. The total number of copies of S_{n-2} in S_n is clearly the number of ways of choosing two letters to fix out of the total of n letters. Recognizing this, we can give the formula for the number of copies of S_{n-k} in S_n: it is given by $_nC_k = (n!)/[k!·(n – k)!]$. Thus we know the basic skeleton of the lattice diagram of S_n.

In addition, for each symmetric subgroup of S_n there is a corresponding alternating subgroup on those letters. Let's consider how this helps us construct a partial lattice diagram of S_4. We know it has a subgroup that is A_4, and from the work of lesson 23 we also know the structure of A_4, all of which is naturally contained in S_4. We now also know that S_4 contains four copies of S_3, which, not being cyclic, we might not have noticed previously. We also know the internal structure of S_3, all of which is contained in S_4, and which organizes the cyclic subgroups we already found. There are other subgroups to be found in S_4 which we have not yet learned about.

It is easy to see how this could get out of hand quickly. In S_6 there will be six copies of S_5, each one with the structure diagrammed here. There will be $_6C_2 = 15$ copies of S_4 distributed among them so that each S_5 contains five of them. Finally there will be $_6C_3 = 20$ copies of S_3 distributed among the fifteen copies of S_4 so that each S_4 contains four of them. To diagram all of the interconnections will create quite a tangle which will be difficult to read in detail. It would not be impossible, but there are many other subgroups which we do not yet know of, and which would need to be included, and which would make the diagram truly unreadable unless spread out over a much wider area. In practice, for very large groups we must make do with a partial lattice diagram combined with a thorough understanding of how the subgroups overlap.

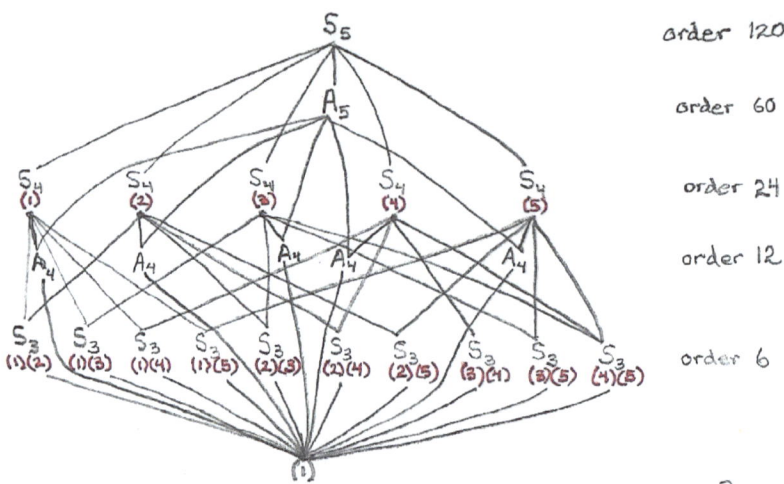

Exercise 74: Finish the lattice diagram of S_4 as far as possible using all that we know about it so far. Use all your work in exercise 72.

Exercise 75: Carefully construct the skeleton of the lattice diagram for S_6 down to the S_3's. Note that there will be 20 copies of S_3, and 15 copies of S_4.

Lesson 24: Dihedral Groups, (odd number of vertices)

We now come to a family of groups which establish a close link between geometry and group theory. In this lesson, we will begin looking at the symmetries of regular polygons. A regular polygon with n sides is defined as the planar figure bounded by n straight lines of equal lengths. A regular polygon with 3 sides is thus an equilateral triangle, a regular polygon with 4 sides is a square, and so on. By a *symmetry* of a figure, we mean a way of moving the figure so that it occupies exactly the same place it occupied before but with the vertices interchanged with each other. In other words, a symmetry is a permutation of the vertices of the polygon such that its shape is preserved.

Not every permutation of the vertices of a polygon are symmetries. For example, draw a square and number the vertices in a clockwise fashion, beginning at any vertex. Now consider the permutation which interchanges the two vertices on the right side, but which leaves the two vertices on the left side fixed. This is not a symmetry because it will force the top and bottom edges to cross each other. The result will not even be a polygon any longer.

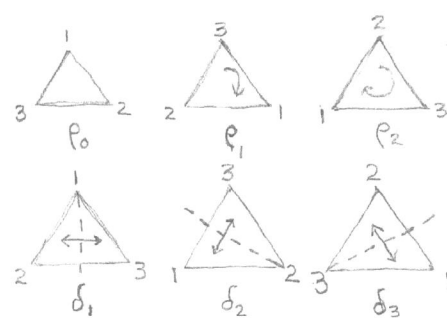

There are two types of symmetries of a polygon: rotations and reflections. In a rotation, the center of the polygon is kept fixed and the vertices are moved rigidly around the center; clockwise is considered the positive direction and counter-clockwise is the negative direction. In a reflection there is an exchange of pairs of vertices across a fixed line through the center of the polygon, which is called the axis of the reflection. Let's begin with the simplest example, the equilateral triangle. All of its symmetries are shown below.

We will use a standard notation for the symmetries of any polygon. Rotations will be denoted by the Greek letter ρ with a subscript; we use the subscript 0 for the identity, which is a rotation through 0°. The subscript 1 is used for the smallest possible rotation that is a symmetry for the polygon, in the case of a triangle, ρ_1 is a clockwise rotation through 120°. We use the multiplicative notation; so ρ_2 is a clockwise rotation through 240°, and since it is the same as two 120° rotations so we know $\rho_2 = \rho_1^2$. In a similar way, since a rotation through 360° is the same as no rotation at all, we know $\rho_0 = \rho_1^3$. Rotations are clockwise, so the inverse of a rotation is a counter-clockwise rotation. Since a clockwise rotation of 240° is the same as a counter-clockwise rotation of 120°, we know that $\rho_2 = \rho_1^{-1}$. Reflections across an axis that passes through a vertex will be denoted by the Greek letter δ. For subscripts we use the number corresponding to the name of the vertex the axis passes through. Reflections clearly are order 2.

Thus there are six symmetries of the equilateral triangle. Geometrically it is obvious that any symmetry followed by any other symmetry must yield a symmetry. In other words, the set of symmetries is closed under the product. Geometrically the product of two symmetries is just one symmetry followed by the other. Labeling the vertices as in the diagram, we can write each of the symmetries as a permutation of the three letters. The product of two symmetries is the same as the product of their corresponding permutations. For the triangle, we have:

$\rho_0 = (1)(2)(3)$ or simply (1) $\rho_1 = (1, 2, 3)$ $\rho_2 = (1, 3, 2)$
$\delta_1 = (1)(2, 3)$ or simply $(2, 3)$ $\delta_2 = (1, 3)$ $\delta_3 = (1, 2)$

Compare these permutations with the diagram to be sure you see the correspondence. Then compare them with the elements of **S₃**. You should see that the group of symmetries of the equilateral triangle is the same group as **S₃**.

We postpone considering the symmetries of the square until the next lesson and skip on to the next polygon with an odd number of vertices, the regular pentagon. This time, labeling the vertices, we will use five letters so this group will be a subgroup of S_5. Again we will have only rotations and reflections; a little thought will show you that there should be five total rotations, including the identity, which is the rotation of 0°. This time the basic rotation is 72°, and so we call this clockwise rotation ρ_1. Then $\rho_2 = \rho_1^2$ is a clockwise rotation through 144°, $\rho_3 = \rho_1^3$ is a clockwise rotation through 216°, $\rho_4 = \rho_1^4$ is a clockwise rotation through 288°. Another rotation of 72° would bring us back to where we started so $\rho_5 = \rho_1^5 = \rho_0$ and we have found all the rotations. In cycle notation we have the following:

$\rho_1 = (1)$ $\rho_1 = (1, 2, 3, 4, 5)$ $\rho_2 = (1, 3, 5, 2, 4)$ $\rho_3 = (1, 4, 2, 5, 3)$ $\rho_4 = (1, 5, 4, 3, 2)$

There are also five reflections, because there are five possible axes of reflection, each one passing through a vertex. In cycle notation the reflections are the following:

$\delta_1 = (2, 5)(3, 4)$ $\delta_2 = (1, 3)(4, 5)$ $\delta_3 = (1, 5)(2, 4)$ $\delta_4 = (1, 2)(3, 5)$ $\delta_5 = (1, 4)(2, 3)$

These are shown in the diagrams below:

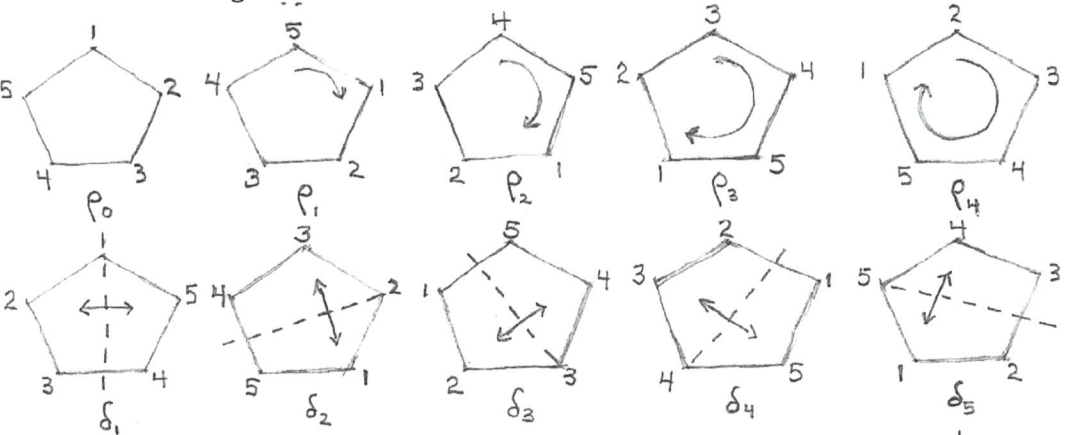

Definition 27: The group of the symmetries of a regular polygon with n vertices is called the **dihedral group on n vertices**. It is denoted by D_n.

If n is the number of vertices, there will be a dihedral group for each n ≥ 3, and thus we have another infinite family of groups. We have presented the elements of D_3, which turned out to be the same group as S_3, and we have presented the elements of D_5, which is a new group for us, a group of order 10, and a subgroup of S_5. The lattice diagrams are given after the exercises.

The regular polygon with n vertices will have 2n symmetries. There will be n rotations, including the identity rotation; and there will be n reflections. The reflections are all of order 2, and the rotations form a cyclic group of order n; thus there are n copies of Z_2 as subgroups of D_n, and one copy of Z_n. Whether there might be other subgroups or not we can't yet say, but we can draw the lattice diagrams for D_3 and D_5. When there are an even number of vertices, the structure of D_n is more complex and so we will consider D_4 in detail in the next lesson.

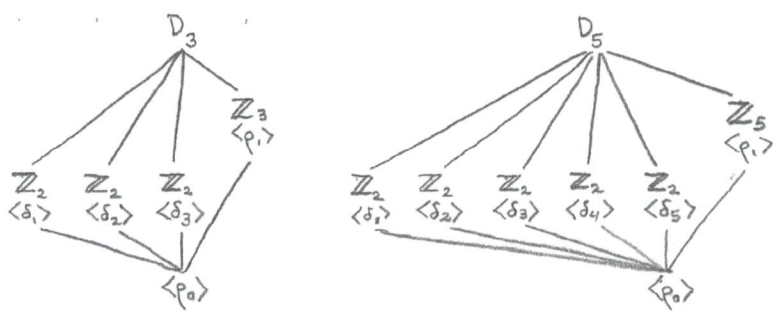

Exercise 76: Make the Cayley table for D_5.

Exercise 77: Find all the elements of D_7. Write them in cycle notation and draw geometric images of how they act on the regular heptagon. Make a Cayley table and a lattice diagram of the result.

Lesson 25: Dihedral Groups, (even number of vertices)

We will continue to look at dihedral groups by considering the regular polygons with an even number of vertices. Because of their higher degree of symmetry, their internal structure is a little more complex than the polygons with an odd number of vertices. So we turn our attention to the symmetries of the square.

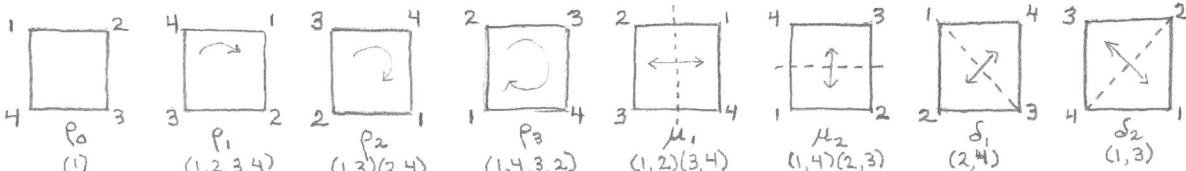

Everything proceeds as for the pentagon or triangle, at first. This time the basic rotation, ρ_1, is a clockwise 90° rotation, ρ_2 is a clockwise 180° rotation, and ρ_3 is a clockwise 270° rotation. There are four reflections, but when we have an even number of vertices, there are two different types of reflections to distinguish: reflections through a diagonal and reflections through a line bisecting a pair of opposite sides. The reflections through diagonals will continue to be denoted by δ with the subscript indicating the vertex with the smaller label contained in the diagonal. Reflections through a bisector of a pair of sides will now be denoted using the Greek letter μ. There is no standard way of assigning the subscript so we will have to specify how it is done each time.

As we expect, there will be one copy of $\mathbf{Z}_4$ from the rotations, and four copies of $\mathbf{Z}_2$ from the reflections, and a fifth copy of $\mathbf{Z}_2$ that is the subgroup of $\mathbf{Z}_4$ generated by the 180° rotation. When a Cayley table is made for $\mathbf{D}_4$, careful observation reveals two more subgroups. It is more enlightening, however, to find these subgroups by geometrical considerations. Polygons with an even number of vertices have a kind of symmetry that polygons with an odd number of vertices don't have, because polygons with an even number of vertices have pairs of reflections whose axes are at right angles to each other. It is always geometrically significant when perpendicularity plays a role in a relationship.

It turns out that when two perpendicular reflections are multiplied, the result is always the 180° rotation, and when that rotation is multiplied by either of the two reflections, the result is always the other reflection. In other words, two perpendicular reflections, the 180° rotation, and the identify form a copy of the Klein 4-group, $\mathbf{V}$. This is not immediately geometrically obvious, so we must give some reason for believing this.

We will first need to divide our argument into two cases: first when n is divisible by 4, and second when n is even but not divisible by 4. In the first case, the two perpendicular reflections will either both be δ's or both will be μ's. In the second case, the two perpendicular reflections will consist of one μ and one δ. To consider the first case, we let n = 4k and use the diagram on the next page.

The two perpendicular δ's are δ_1 and δ_{k+1} as shown in the diagram on the next page. We can use the diagram to express the two reflections in cycle notation leaving the number of vertices an arbitrary multiple of 4:

$\delta_1 = (2, 4k)(3, 4k-1)(4, 4k-2)...(m, 4k-m+2)...(2k-1, 2k+3)(2k, 2k+2)$

As a check, we know we have the right pattern because vertex m on the right will reflect over to vertex $4k - m + 2$ on the left. Thus when m = 2k − 1 it should be reflected to

$4k - m + 2 = 4k - (2k - 1) + 2 = 2k + 3$

which is what we expect from the diagram. Note that both 1 and 2k + 1 are fixed. In the same way it is relatively easy to see that

$\delta_{k+1} = (k, k+2)(k-1, k+3)...(k-m, k+m+2)...(1, 2k+1)(4k, 2k+2)(4k-j, 2k+j+2)...(3k+2, 3k)$

and you can easily check the correspondence. Now if we multiply these two transpositions in either order we will get: $(1, 2k + 1)(2, 2k + 2)...(k, 3k)...(2k, 4k) = \rho_{2k}$ which is the 180° rotation for a polygon with 4k vertices. Follow a few of the vertices through the multiplication to convince yourself. For example, in δ_{k+1} 3k is mapped to 3k + 2. To see how 3k + 2 is mapped by δ_1 we must choose m so that $4k - m + 2 = 3k + 2$, which gives m = k. Thus we see that 3k + 2 is mapped to k. Thus the product maps 3k to k as expected.

In the same diagram the perpendicular μ reflections are shown in red. Expressing these in cycle notation and then multiplying them to get ρ_{2k} is left as an exercise. The other case involves an even number of vertices not divisible by 4, 4k + 2 vertices. In this case the perpendicular reflections are δ_1 and for μ_{k+1} as shown in the right hand diagram. Expressing these as cycles and then multiplying them is also left as an exercise.

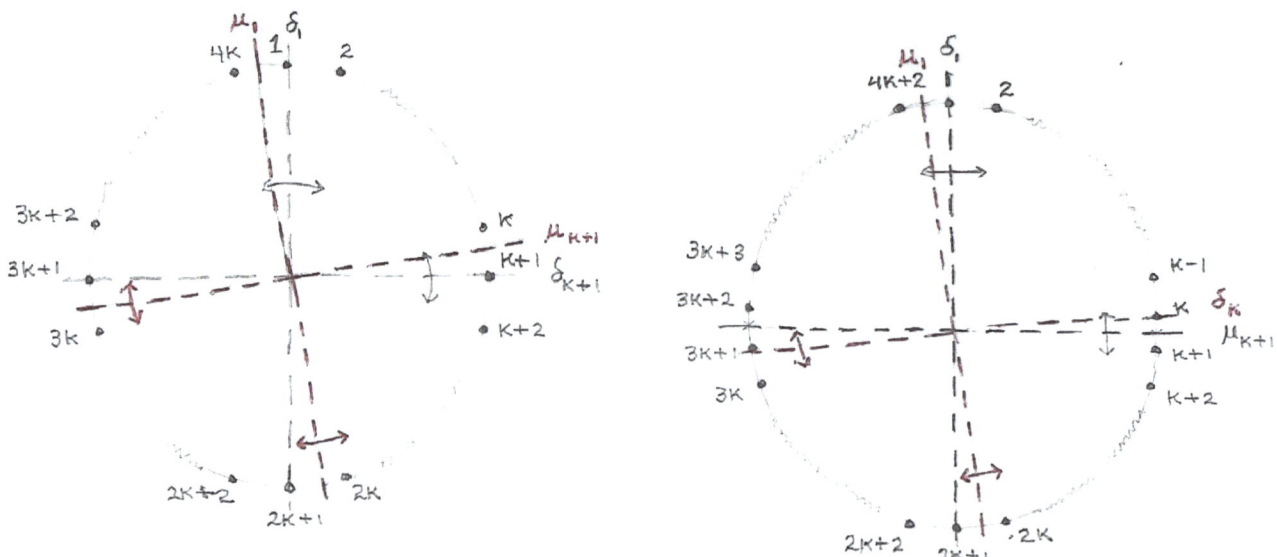

Let's conclude this lesson by examining **D₆** in some detail. There is an additional level of complexity in this group that we need to consider. The elements of this group are illustrated below and are given in cycle notation. I have paired the perpendicular reflections to show the three different copies of **V** we expect to find, namely $\{\rho_0, \rho_3, \delta_1, \mu_3\}$, $\{\rho_0, \rho_3, \delta_2, \mu_1\}$, and $\{\rho_0, \rho_3, \delta_3, \mu_2\}$. But there are other subgroups present that we should know, with a little thought, must be present. The permutations we are considering preserve the regular hexagon and a hexagon has four equilateral triangles that can be inscribed in it. It may be that these triangles are preserved by some of the permutations of the hexagon, and if so then the subset of permutations that preserve a triangle will corporately form a copy of **D₃**. In fact, a little thinking shows that there are two pairs of triangles preserved by two different subsets of permutations. The permutations $\delta_1, \delta_2, \delta_3, \rho_0, \rho_2$, and ρ_4 preserve on pair of triangles; and $\mu_1, \mu_2, \mu_3, \rho_0, \rho_2$, and ρ_4 preserve the other pair, as illustrated in the diagram on the next page, with one pair of triangles marked in red and the other pair marked in green.

This gives a complete picture of the structure of **D₆**, one of the most intricate groups that we have studied so far. Intricate it may be, but the group itself is relatively straight-forward and easy to understand. The complete lattice diagram is shown on the next page.

Exercise 78: Complete the other cases mentioned above to show that the product of two perpendicular reflections is the 180° rotation.

Exercise 79: Construct the Cayley tables for **D₄** and **D₆**.

Exercise 80: Construct the Cayley tables and lattice diagrams for **D₈** and **D₁₀**.

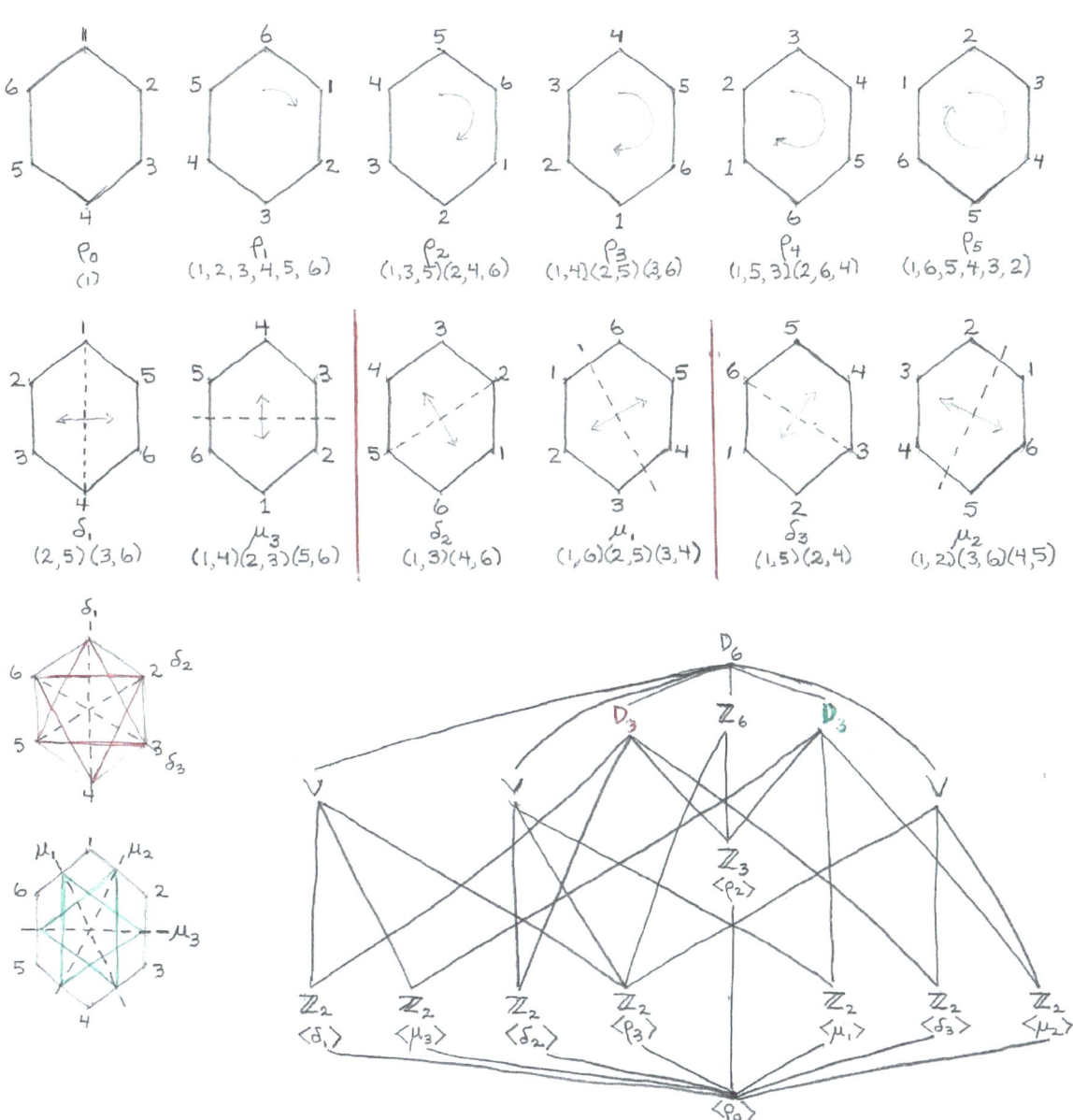

Lesson 26: Dihedral Subgroups of S_n

Since D_n is a group of permutations of the n vertices of a regular polygon, it is clearly also a subgroup of S_n, a subgroup we have not yet put into the lattice diagram because we didn't know it existed. Now that we do know it exists, how many copies of D_n should we expect to find in S_n?

The standard copy of D_n is obtained by assigning the letters to the vertices of an n-gon consecutively moving around the figure clockwise. These are then permuted by rotations and by reflections to give a total of 2n permutations out of the n! possible permutations. In order to find how many copies of D_n are contained in S_n, we need only look at the cyclic group of rotations within D_n. Each labeling of the vertices will determine a dihedral group, but there will be duplication. Once the vertices are labeled, the basic rotation in <u>that</u> group will be written as an n-cycle. Two different labelings will determine the same subgroup if the basic rotation of one labeling is a power of the basic rotation of the other; in other words, it they are both generators of the same copy of the cyclic group of rotations, Z_n. We know that the cyclic group Z_n has $\varphi(n)$ distinct generators. We need only know how many n-cycles there are in S_n.

In combinatorics, the assignment of letters to vertices is called a circular permutation. So we are asking how many circular permutations of n letters there are. The answer, which we will take as answered by combinatorics, is $(n-1)!$. Since $\varphi(n)$ n-cycles determine a single dihedral group we see that the number of copies of D_n in S_n equals $(n-1)!/\varphi(n)$.

Since we know S_3 is the same group as D_3 we know there is only one copy of D_3 in it. By the formula we just derived, we compute that S_3 should have $(3-1)!/\varphi(3) = 2/2 = 1$.

S_4 of course is more complicated. We expect there to be $(4-1)!/\varphi(4) = 3!/2 = 3$ copies of D_4. You will be asked to find these three copies of D_4 in the exercises.

The example we will do in some detail now is to find all the copies of D_5 in S_5. We expect there are $(5-1)!/\varphi(5) = 4!/4 = 6$ of them. There are four 5-cycles generating the rotation group in each copy of D_5, and a total of 24 5-cycles. All of the 5-cycles can be listed by being organized and careful in how we search. Since the first letter given in a cycle is arbitrary we may as well take it as the letter 1. The second vertex must be labeled 2, or 3, or 4, or 5, so we begin our search with four columns. The third vertex must be labeled one of the three remaining letters, so each of the four columns gets split into three sub-columns. Finally there are only two possible letters left for the fourth vertex, and the fifth vertex is forced to be the lone remaining letter. Thus in each of our twelve sub-columns we can write two 5-cycles. Once this is done we have accounted for all the expected 5-cycles.

To see which 5-cycles are grouped together into a single D_5, simply choose any 5-cycle and compute the subgroup it generates and put all those 5-cycles together. Then choose one of the remaining 5-cycles and do the same thing. Repeat this until they are all sorted into a copy of D_5. In each group choose any of the four 5-cycles as the basic one for labeling a pentagon.

Once the vertices are labeled, it is a simple matter to compute the permutations that represent the five reflections in each group. Note that distinct pentagons, distinct copies of D_5 may share some reflections; that is, distinct copies of D_5 may share copies of Z_2. This sharing of reflections must be indicated on the lattice diagram, of course, and may lead to a considerable tangle. On the next page I have given the six pentagons and the elements of each copy of D_5. I have tried to indicate through colors and different types of underlining when various reflections are repeated.

The lattice diagram previously given for S_5 was only a partial diagram. It included only the alternating and symmetric subgroups, basically the top part of the diagram. Inserting these dihedral subgroups will greatly complicate it. Care must be taken to get all the interconnections right. Notice that every element of every one of these copies of D_5 is an even permutation, so all of these subgroups are contained in the alternating subgroup A_5.

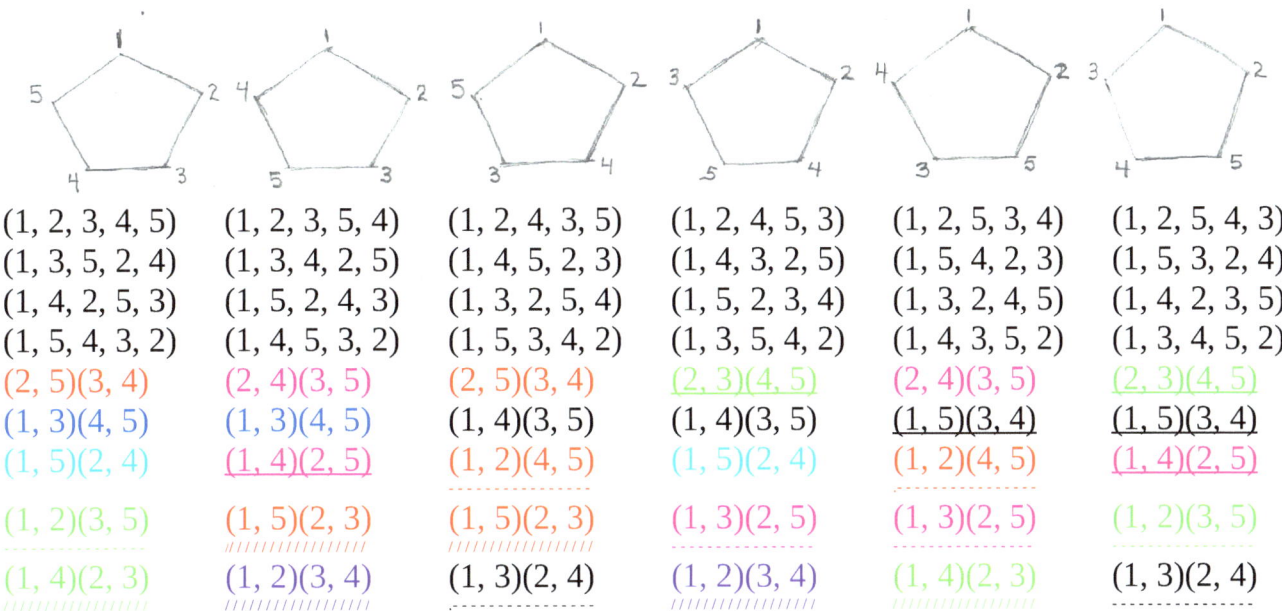

(1, 2, 3, 4, 5)	(1, 2, 3, 5, 4)	(1, 2, 4, 3, 5)	(1, 2, 4, 5, 3)	(1, 2, 5, 3, 4)	(1, 2, 5, 4, 3)
(1, 3, 5, 2, 4)	(1, 3, 4, 2, 5)	(1, 4, 5, 2, 3)	(1, 4, 3, 2, 5)	(1, 5, 4, 2, 3)	(1, 5, 3, 2, 4)
(1, 4, 2, 5, 3)	(1, 5, 2, 4, 3)	(1, 3, 2, 5, 4)	(1, 5, 2, 3, 4)	(1, 3, 2, 4, 5)	(1, 4, 2, 3, 5)
(1, 5, 4, 3, 2)	(1, 4, 5, 3, 2)	(1, 5, 3, 4, 2)	(1, 3, 5, 4, 2)	(1, 4, 3, 5, 2)	(1, 3, 4, 5, 2)
(2, 5)(3, 4)	(2, 4)(3, 5)	(2, 5)(3, 4)	(2, 3)(4, 5)	(2, 4)(3, 5)	(2, 3)(4, 5)
(1, 3)(4, 5)	(1, 3)(4, 5)	(1, 4)(3, 5)	(1, 4)(3, 5)	(1, 5)(3, 4)	(1, 5)(3, 4)
(1, 5)(2, 4)	(1, 4)(2, 5)	(1, 2)(4, 5)	(1, 5)(2, 4)	(1, 2)(4, 5)	(1, 4)(2, 5)
(1, 2)(3, 5)	(1, 5)(2, 3)	(1, 5)(2, 3)	(1, 3)(2, 5)	(1, 3)(2, 5)	(1, 2)(3, 5)
(1, 4)(2, 3)	(1, 2)(3, 4)	(1, 3)(2, 4)	(1, 2)(3, 4)	(1, 4)(2, 3)	(1, 3)(2, 4)

Exercise 81: Redraw the lattice diagram of S_5 to include all the copies of D_5 and the subgroups of each D_5. Be careful to show correctly how all subgroups are shared among them. (Do not try to include any copies of D_4, which must obviously be present as well.)

Exercise 82: How many copies of D_7 are contained in S_7? Will the elements be even or odd permutations or some of each?

Exercise 83: Determine completely all the copies of D_4 that are contained in S_4 and give a complete lattice diagram of S_4 which includes them.

Lesson 27: The Symmetries of the Cube

We will now consider what happens in three dimensions. Symmetries in two dimensions are relatively easy, but in three dimensions new and unexpected things happen. Here we will examine in detail one of the simplest three dimensional objects, the cube, just to get a glimpse of some of the difficulties that may arise. Naturally, going to even higher dimensions will be much more difficult, if only because our ability to visualize higher dimensional objects is severely limited.

We can begin as we would in the two dimensional case, with what is familiar and carries over. First let's look at the reflections. As previously, there are two different types of reflections. The plane of the reflection may bisect edges of the cube, corresponding to the μ reflections in the two dimensional case. The cube has three of these – the up/down reflection, the left/right reflection, and the front/back reflection. All the diagrams are placed at the end of this lesson for reference. Labeling the vertices as in those diagrams we have the following:

$\mu_1 = (1, 2)(3, 4)(5, 6)(7, 8)$
$\mu_2 = (1, 4)(2, 3)(5, 8)(6, 7)$
$\mu_3 = (1, 5)(2, 6)(3, 7)(4, 8)$

In the second type of reflection, the plane of the reflection contains two diagonally opposite edges, corresponding to the δ reflections in the two dimensional case. There are twelve edges in a cube, so there are six opposite pairs of edges. This gives us:

$\delta_1 = (4, 5)(3, 6)$ $\qquad\delta_2 = (1, 6)(4, 7)$ $\qquad\delta_3 = (1, 8)(2, 7)$
$\delta_4 = (2, 5)(3, 8)$ $\qquad\delta_5 = (2, 4)(6, 8)$ $\qquad\delta_6 = (1, 3)(5, 7)$

These are all order two elements, and thus we have ten elements of our group, including the identity. The rotations are more complicated than in two dimensions. There are three distinct types of rotations. In the first, the axis of rotation intersects the centers of two opposite faces. Since there are six faces available, there are three distinct rotations of this type, each generated by a 90° rotation. These rotations each have order 4. We will call these rotations ρ rotations:

$\rho_1 = (1, 2, 3, 4)(5, 6, 7, 8)$ $\quad\rho_1^2 = (1, 3)(2, 4)(5, 7)(6, 8)$ $\quad\rho_1^3 = (1, 4, 3, 2)(5, 8, 7, 6)$
$\rho_2 = (1, 4, 8, 5)(2, 3, 7, 6)$ $\quad\rho_2^2 = (1, 8)(4, 5)(2, 7)(3, 6)$ $\quad\rho_2^3 = (1, 5, 8, 4)(2, 6, 7, 3)$
$\rho_3 = (1, 2, 6, 5)(3, 7, 8, 4)$ $\quad\rho_3^3 = (1, 6)(2, 5)(3, 8)(7, 4)$ $\quad\rho_3^3 = (1, 5, 6, 2)(3, 4, 8, 7)$

We now have a total of 19 elements in this group, including the identity.

In the second type of rotation, the axis passes through the center of a pair of opposite edges, making six rotations of this type. Each one is a 180° rotation and so these are all order 2. We will use the letter σ for this type of rotation. This gives us:

$\sigma_1 = (1, 2)(3, 5)(4, 6)(7, 8)$ $\qquad\sigma_2 = (1, 7)(4, 6)(2, 3)(5, 8)$ $\qquad\sigma_3 = (1, 7)(2, 8)(3, 4)(5, 6)$
$\sigma_4 = (1, 4)(2, 8)(3, 5)(6, 7)$ $\qquad\sigma_5 = (1, 7)(2, 6)(3, 5)(4, 8)$ $\qquad\sigma_6 = (1, 5)(2, 8)(3, 7)(4, 6)$

We now have a total of 25 elements.

The third type of rotation is the one most often over-looked by the novice. The axis of rotation may pass through two diagonally opposite vertices. In this case, the generating rotation is 120° and each of the four possible generators has order 3. We will use the letter τ for this type of rotation. Thus we have:

$\tau_1 = (1, 6, 8)(2, 7, 4)$ $\qquad\tau_1^2 = (1, 8, 6)(2, 4, 7)$
$\tau_2 = (1, 3, 6)(4, 7, 5)$ $\qquad\tau_2^2 = (1, 6, 3)(4, 5, 7)$
$\tau_3 = (1, 3, 8)(2, 7, 5)$ $\qquad\tau_3^2 = (1, 8, 3)(2, 5, 7)$
$\tau_4 = (2, 5, 4)(3, 6, 8)$ $\qquad\tau_4^2 = (2, 4, 5)(3, 8, 6)$

This brings our total number of elements to 33.

It is here that something, possibly unexpected happens – a new symmetry that is neither a reflection nor a rotation. We discover it in a routine manner. The group we are constructing contains these 33 elements, and all their possible products. So let's consider the product

$\mu_3 \mu_2 \mu_1$ = (1, 5)(2, 6)(3, 7)(4, 8)(1, 4)(2, 3)(5, 8)(6, 7)(1, 2)(3, 4)(5, 6)(7, 8)
= (1, 7)(2, 8)(3, 5)(4, 6)

Each vertex goes to the opposite vertex. This is a new element for the group, the 34th, and a new kind of symmetry which we will call **inversion**. We will name it with the letter λ.

At this point we must check for other products which yield new elements of this group. It is enough to try a representative product from the different types of elements. To shorten the discussion a bit, what we show here are the products of λ with the τ's and then the products of λ with the σ's. It will turn out that these products give us all the unaccounted for elements of the group, though you must check other products (an exercise).

The product of λ with a τ will be denoted by the letter ζ:

$\zeta_1 = \tau_1 \lambda$ = (1, 6, 8)(2, 7, 4)(1, 7)(2, 8)(3, 5)(4, 6) = (1, 4, 8, 7, 6, 2)(3, 5)

Surprisingly, this is an element of order 6 so we need to consider its powers:

ζ_1^2 = (1, 8, 6)(2, 4, 7) = τ_1^2. Hence we know $\zeta_1^4 = \tau_1^4 = \tau_1$
ζ_1^3 = (1, 7)(4, 6)(2, 8)(3, 5) = λ and of course $\zeta_1^5 = \zeta_1^{-1}$ = (1, 2, 6, 7, 8, 4)(3, 5)

Thus there are only 2 new elements. The same pattern will occur for the other τ's:

$\zeta_2 = \tau_2 \lambda$ = (1, 3, 6)(4, 7, 5)(1, 7)(2, 8)(3, 5)(4, 6) = (1, 5, 6, 7, 3, 4)(2, 8)
ζ_2^2 = (1, 6, 3)(4, 5, 7) = τ_2^2 and hence $\zeta_2^4 = \tau_2$
ζ_2^3 = (1, 7)(3, 5)(4, 6)(2, 8) = λ and $\zeta_2^5 = \zeta_2^{-1}$ = (1, 4, 3, 7, 6, 2)(3, 5)

There will be a total of 4 new elements from the other two products, as you can easily check:

$\zeta_3 = \tau_3 \lambda$ = (1, 3, 8)(2, 7, 5)(1, 7)(2, 8)(3, 5)(4, 6) = (1, 5, 8, 7, 3, 2)(4, 6)
ζ_3^{-1} = (1, 2, 3, 7, 8, 5)(4, 6)
$\zeta_4 = \tau_4 \lambda$ = (2, 5, 4)(3, 6, 8)(1, 7)(2, 8)(3, 5)(4, 6) = (2, 6, 5, 8, 4, 3)(1, 7)
ζ_4^{-1} = (2, 3, 4, 8, 5, 6)(1, 7)

Study the pictures of these symmetries until you begin to see the pattern. These are symmetries we could not have detected easily except for our knowledge of groups. We now have a total of 42 elements in this group.

It turns out, as you can check, that the products of λ with the ρ's give the last of the symmetries of the cube. We will name these new products using the letter ξ. In brief the results are:

$\xi_1 = \rho_1 \lambda$ = (1, 2, 3, 4)(5, 6, 7, 8)(1, 7)(2, 8)(3, 5)(4, 6) = (1, 8, 3, 6)(2, 5, 4, 7)
and we easily see that ξ_1^{-1} = (1, 6, 3, 8)(2, 7, 4, 5)
and that $\xi_1^2 = \rho_1^2$
$\xi_2 = \rho_2 \lambda$ = (1, 4, 8, 5)(2, 3, 7, 6)(1, 7)(2, 8)(3, 5)(4, 6) = (1, 6, 8, 3)(2, 5, 7, 4)
and you can verify that ξ_2^{-1} = (1, 3, 8, 6)(2, 4, 7, 5)
and that $\xi_2^2 = \rho_2^2$
$\xi_3 = \rho_3 \lambda$ = (1, 2, 6, 5)(3, 7, 8, 4)(1, 7)(2, 8)(3, 5)(4, 6) = (1, 8, 6, 3)(2, 4, 5, 7)
and you can verify that ξ_3^{-1} = (1, 3, 6, 8)(2, 7, 5, 4)
and that $\xi_3^2 = \rho_3^2$

These 48 elements are all the symmetries of the cube, and here we can see the value and power of taking the abstract group theoretic approach to symmetries. Using the computations above you should have no trouble in filling in a Cayley table for this group, and constructing the major part of the lattice diagram, which of course is an exercise. This group is denoted by $\mathbf{O^*_{48}}$. The $\mathbf{O}$ is used because this is connected to the octahedron's symmetries as you will be asked to show. The 48 indicates the order of the group.

This is part of a small family of groups which we cannot discuss at this point, but they will be quite important in more advanced studies. The study of the symmetries of the regular polyhedra played an important role in the development of the group concept in the late 19th century. We will have occasion in future parts of this series to return to this topic

Only the generators of the various symmetries are diagrammed below, so not all the elements of the group are shown.

Exercise 84: Construct the Cayley table, and as much of the lattice diagram as possible, for the group of symmetries of the cube.

Exercise 85: Follow the process for the cube to find the symmetries for the octahedron. Once you have found all the symmetries, can you identify the group as one you have seen before?

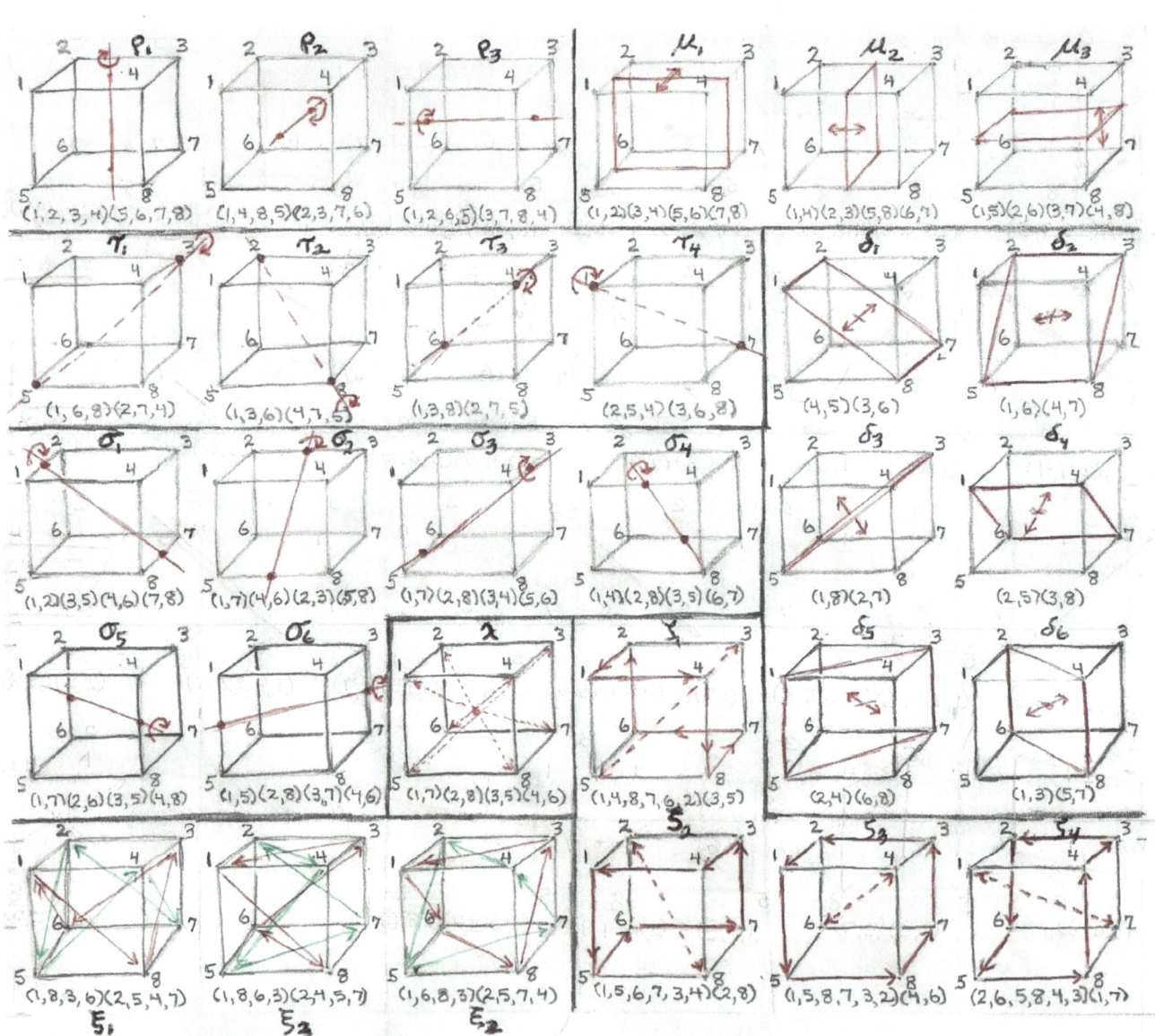

Lesson 28: Group Presentations

Of all groups, only cyclic groups are generated by a single element; that's their definition. All other groups, and in particular the dihedral groups, require two or more elements to generate them. Now it is time to consider how to choose the generators, how to manage them, how to use them to gain insight into the group itself. What motivates us here is that we need more tools we can use to analyze the structure of a group. The main tool we have at the moment is the lattice diagram, which is extremely difficult to construct for even moderately large groups. It would be very useful if we had a more algebraic tool as well, and one which could be more easily used. Finding the generators of a group is the obvious place to start. Since they are the generators, we would expect them to contain *all* the algebraic information about the group.

Dihedral groups are a good place to begin this discussion, rather than the symmetric or the alternating groups, because dihedral groups are simpler and need only two generators. We know that D_n includes n rotations (where we take the identity element as a rotation of 0° or of 360°). The other n elements are all reflections. The n rotations together form a cyclic subgroup, Z_n, which is generated by the smallest of the rotations. Clearly we must look for another generator among the reflections. It is not immediately clear that a single reflection, along with the generating rotation, will be enough to determine the group, or whether we will need more of them. Furthermore, it is not immediately clear which reflection we should choose to begin. Are they all equally useful, or are only a few particular reflections capable of generating the group? As it turns out, the reflections are all equally capable of use as the additional generator.

We will approach the problem abstractly. Let a represent a fixed reflection; let r represent the minimal rotation, a generator of the subgroup of all rotations. Then I claim D_n = < a, r >. Clearly, this is insufficient to explain D_n. We also need to show the orders of the generators, and how the group operation works with the generators. In other words we must know how to calculate products of powers of the two generators. At the very least, we should be able to compute the Cayley table for the group using only the generators and a few equations relating them.

For D_n - for any group, actually - we will first specify the orders of the generators by equations. In this case the required equations are $a^2 = 1$ and $r^n = 1$. We will make the assumption that n is the order of r; i.e. that no power of r smaller than n makes it equal 1. This is still not enough since it does not tell us how to multiply a with r. We look for an algebraic equation relating a and r. Among other things, it must tell us how to switch the order of the multiplication. For the dihedral group we can specify how to multiply the two generators with this equation: $ar = r^{-1}a$. We could have stated it as rar = a, which is equivalent, but the main use of the relation will be to reverse the order of the generators and for that the first form is more useful.

Next we must choose names, somehow, for all the elements of D_n. Half of them are already named: $1, r, r^2, \ldots, r^{n-1}$. The other half of the elements must be products of a and r. The obvious choice to try is $a, ar, ar^2, \ldots, ar^{n-1}$. If these are suitable, if there are no duplicates among them, then this would give us all the 2n elements in D_n. Hence we need only show that the elements $1, r, r^2, \ldots, r^{n-1}, a, ar, ar^2, \ldots, ar^{n-1}$ are all distinct.

We can show this by contradiction. It is easy to see that ar^m cannot equal any r^k. If it were, then clearly we would have $a = r^{k-m}$, which contradicts our assumption that a is not a rotation. Similarly, no two powers of r can equal each other also by assumption that they are a list of all the element of the cyclic subgroup. We need only check if ar^m can be equal to ar^k when m and k are both less than n. We may assume that m < k without any loss of generality. So beginning with $ar^m = ar^k$, first multiply through on the left by the inverse of a which is a. This will give us $r^m = r^k$. Since both m and k are both smaller than n, we know this is impossible in the cyclic group < r > of order n, unless m=k. Thus we have shown that the 2n elements in our list are all distinct and we know we have named every element of D_n.

The next question is: is the relation $ar = r^{-1}a$ adequate for making every possible computation in D_n? To do this the relation must enable us to switch the order of multiplication in any product. To this end, first note that
$$ar = r^{-1}a \Rightarrow ara = r^{-1} \Rightarrow ra = ar^{-1}$$
We can now make the following computation:
$$r^k(ar^m) = (r^k a)r^m = (r^{k-1}ra)r^m = (r^{k-1}ar^{-1})r^m = (r^{k-1}a)r^{-1}r^m = (r^{k-1}a)r^{m-1}$$
Thus we can move all k of the r's from the left of the a over to the right of the a one at a time. Each time we move an r over we reduce the power of r by 1 on each side of the equation. Eventually we can obtain a named element in our list as the result:
$$r^k(ar^m) = ar^{m-k}$$
In this same way we can compute:
$$ar^k(ar^m) = r^{m-k}$$
This gives us a way to compute any possible product of elements of D_n all in terms of the elements in the list.

If we use these formulas we can fill in the Cayley table for D_4 using only the relations:
$$a^2 = 1,\ r^4 = 1,\ \text{and } ar = r^{-1}a = r^3 a$$
noting that the last relation is the same as $ra = ar^3$.

$1 = r^0$	r	r^2	r^3	a	ar	ar^2	ar^3
r	r^2	r^3	1	ar^3	a	ar	ar^2
r^2	r^3	1	r	ar^2	ar^3	a	ar
r^3	1	r	r^2	ar	ar^2	ar^3	a
a	ar	ar^2	ar^3	1	r	r^2	r^3
ar	ar^2	ar^3	a	r^3	1	r	r^2
ar^2	ar^3	a	ar	r^2	r^3	1	r
ar^3	a	ar	ar^2	r	r^2	r^3	1

Compare this with the Cayley table for D_4 that you computed as an exercise. The r may correspond to either the 90° clockwise rotation or the 270° clockwise rotation; it is not specified in the presentation. The a could correspond to any of the four reflections; again, it is not specified by the presentation. Does the choice of the two specific generators make a difference in the two Cayley tables, or do we have to rearrange the rows or columns to make them match?

Definition 28: A list of an adequate set of generators, together with an adequate set of relations between them, is called a **presentation** for the group.

We will write a presentation by using a carat to begin, then listing the chosen generators, then a colon, then a list of the necessary relations between the generators, and then a closing carat. So the above presentation for D_n is written this way:
$$D_n = <a, r : a^2 = 1,\ r^n = 1,\ ar = r^{-1}a>$$
Note that in a presentation, since it is presumed to present a group, we automatically take the associative law for granted.

The presentation for a particular group is usually not unique. There is almost always a choice of which elements we take for the generators; and sometimes different choices of generators require different equations to express the relationships between them. When there is a choice of presentations, one may be more convenient than another depending on the problem to be solved. In the case of the dihedral groups there is a second presentation that gives useful information that the preceding presentation does not give. As an alternative to the presentation given above, we can use two reflections to generate the entire group. This gives a presentation of the following form:

$$\mathbf{D_n} = \ <a, b : a^2 = b^2 = 1, (ab)^n = 1>$$

While we can derive the entire dihedral group using only two reflections, we are not free to choose just any two reflections we wish. Not every possible pair will generate $\mathbf{D_n}$. For example, choosing two perpendicular reflections will give the 180° rotation as the product instead of a generating rotation. We will discuss the choice of which reflections work in the next lesson. At any rate, one disadvantage of this second presentation is that we must take more care choosing the generators. Some care is always required; in the original presentation above, we still had to be careful to choose a generating rotation for the subgroup of rotations.

But there is a substantial insight to be gained by this second presentation. It gives us a huge help in deciphering the structure of a group G: it gives us a test to detect the presence of dihedral subgroups. First, find all the cyclic groups generated by the elements of G. Collect all the elements of order 2. If there are more than one, then take the order 2 elements in pairs, say a and b, and find the order of their products ab. The products will never equal the identity since that would mean the two elements were the same. If the product also has order 2 then a, b, and ab together form the Klein 4-group. If a product has order n > 2, then a, and b generate a dihedral group, namely $\mathbf{D_n}$, as the second presentation shows. This gives us an easy way to determine a second layer of subgroups above the cyclic subgroups. It does not, of course, tell us all the subgroups by a long shot, but it reveals a layer of the structure we would otherwise not have easily noticed as.

Exercise 86: Show that the presentation $<a, b : a^2 = b^2 = 1, (ab)^n = 1>$ produces the same group as the presentation $<a, r : a^2 = 1, r^n = 1, ar = r^{-1}a>$.

Exercise 87: Give the Cayley table for $\mathbf{D_4}$ using only the presentation $<a, b : a^2 = b^2 = 1, (ab)^4 = 1>$.

Lesson 29: The Quaternion Group

So far we have presented four families of finite groups (Z_n, S_n, A_n and D_n), and one family that is really more of an association than a family (U_n). We began with the cyclic groups, Z_n, of order n, characterized by the fact that each one is generated by a single element. These are the simplest of all groups and are all Abelian. All groups are composed of cyclic groups like all chemicals are composed of atoms of various elements. When analyzing the structure of a new group the first step is to find all of its cyclic subgroups and map out their relationships with each other.

Then we discussed the symmetric groups, S_n, of order n!. These are groups of permutations, i.e. of bijections, on a set of n letters. They are the opposite of the cyclic groups in nature. If the cyclic groups are like atoms, then the symmetric groups are like galaxies, the most complicated groups into which all other finite groups fit. Closely related to them are the alternating groups, A_n, of order ½·n! consisting of the even permutations. Finally we discuss the dihedral groups, D_n, of order 2n, defined geometrically as the symmetries of regular polygons. We also spent a lot of time discussing a less coherent collection of groups, the group of units of the integers mod n under multiplication, U_n. The exact nature of these groups will be made more clear in the next book on groups, when we have more algebraic tools at our disposal.

There are many other families of finite groups which we are not prepared to discuss until we have more algebraic sophistication. However, there is one more family of groups that we can discuss at this point, both because of their importance and because they are best defined using their group presentations. The first and smallest one of this family is called **the quaternion group**. It could be defined without mentioning its presentation, but since it is the purpose of this lesson and the next to gain skill at using presentations we will define it this way.

The quaternion group is more abstract in nature. All the preceding families of groups could engage the intuition on some level, as numbers mod n or as rearrangements of a set or as symmetries of a geometrical object; but the quaternions appeal directly to the imagination only when we arrive at a more advanced understanding of algebra. The quaternion group will be denoted by Q_8. Here the subscript denotes the order of the group. It was natural to use the subscript of Z_n to denote the order of the cyclic group, but it was more natural to use the subscripts of S_n and A_n to denote the number of letters being permuted, rather than the order; and it was more natural to use the subscript of D_n to denote the number of vertices of the polygon whose symmetries it represents (though most other authors disagree and use the subscript of the dihedral group to denote its order; be aware of that when you read other texts!)

The quaternion group can be presented with only two generators this way:
$$Q_8 = \langle i, j : i^4 = j^4 = 1, i^2 = j^2, ij = ji^{-1} \rangle$$
There are other ways we could have given the relations. It is more standard to give the third relation as $j^{-1}ij = i^{-1}$ for reasons we will get to in the second volume of Modern Algebra. We could also have given the third relation as $ij = ji^3$ but it is usually more helpful to be explicit when an inverse element is involved.

Now let's consider how to name the eight elements of the group. Because $i^4 = 1$, we may be reminded of the imaginary unit $\breve{i}$ in the complex numbers. We will take advantage of this similarity and use $\breve{i}$ instead of i, and $\breve{j}$ instead of j. Further, we will use -1 to represent $\breve{i}^2 = \breve{j}^2$. This makes it natural to use $-\breve{i}$ in place of $\breve{i}^{-1} = \breve{i}^3$ and $-\breve{j}$ instead of $\breve{j}^{-1} = \breve{j}^3$. This gives us names for six out of the eight elements we are expecting: 1, -1, $\breve{i}$, $-\breve{i}$, $\breve{j}$, and $-\breve{j}$. These new names for the elements also change the appearance of the relations. We now have:
$$Q_8 = \langle \breve{i}, \breve{j} : \breve{i}^4 = \breve{j}^4 = 1, \breve{i}^2 = \breve{j}^2 = -1, \breve{i}\breve{j} = -\breve{j}\breve{i} \rangle$$
There are two more elements we expect to need to name. First compute the order of $\breve{i}\breve{j}$:

$(ĭĵ)^2 = ĭĵ\ ĭĵ = ĭĵ(ĵĭ^{-1}) = ĭ(ĵ^2)ĭ^{-1} = ĭ^2\ ĭ^{-1} = ĭ^2 = -1$
$(ĭĵ)^3 = ĭĵ\ (ĭĵ)^2 = ĭĵ\ ĵ^2 = ĵĭ^{-1}ĵ^2 = ĵĭ^{-1}ĭ^2 = ĵĭ$
also $ĭĵ(ĭĵ)^2 = ĭĵ(-1)$

Clearly then $ĭĵ \neq ĵĭ$. If we continue our pattern of naming the elements, we should choose $ǩ = ĭĵ$ and $-ǩ = ĵĭ$. We have not yet proven that the collection of elements $±1$, $±ĭ$, $±ĵ$, and $±ǩ$ do form a group. In the exercises you will be asked to prove this is a group by the awkward means of computing the Cayley table.

The quaternions are now the third group of order 8 that we have encountered so we are in a position to better compare the possible group structures that can exist for a particular order. In fact there are two more groups of order 8 but we are not ready to discuss them at this point. Below are the lattice diagrams for the three groups we know: **Z_8, D_4,** and **Q_8**. Study them until you get a clearer picture of the internal structures involved. Shortly we will develop another kind of diagram of the group structure which will be helpful in other ways.

Exercise 88: Construct the Cayley table for **Q_8**, using only the presentation.

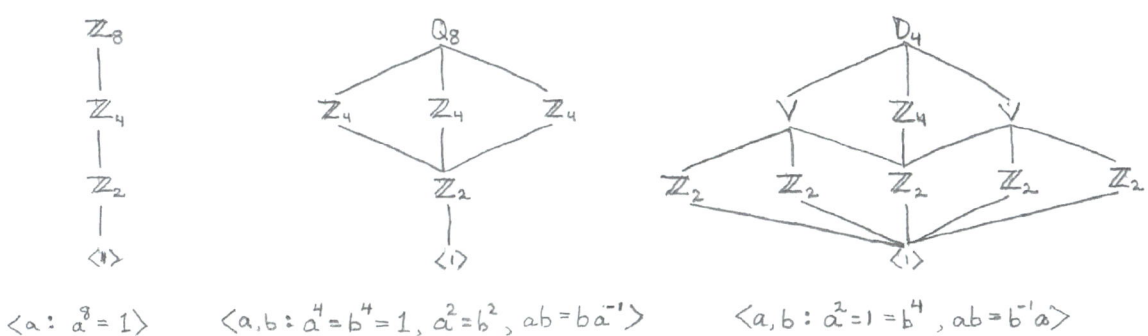

Lesson 30: The Generalized Quaternions

We will use a presentation to generalize the quaternion group, to define an infinite family of groups which are all similar to each other in certain respects and which has the quaternion group as its simplest member. The difficulty is that there is no clear path that leads to this generalization. What is it that is essential to the quaternion group that should be duplicated in the family we seek? And once we know the essential quality we wish to preserve, how can we do it? These are critical questions, but they can not be clearly explained at this point. For example, one essential property of the quaternion group is that it contains exactly one subgroup of order 2 and that is something we will insist on in the generalized quaternions we will define. Every generalized quaternion group has exactly one element of order 2. The reason why this is an important property for us can not be made clear for many lessons yet. Meanwhile

Definition 29: The **generalized quaternion groups** are given by the presentation:
$$Q_{2^{\wedge}(n+1)} = < x, y : x^{2^{\wedge}n} = y^4 = 1; x^{2^{\wedge}(n-1)} = y^2; xy = yx^{-1} >$$

There are several things you should notice immediately from this presentation. First, the subscript represents the order of the group, so a generalized quaternion group always has a power of 2 for its order. The smallest member of this family, the quaternion group Q_8, is found by setting n = 2. It is an exercise to determine what group results by setting n = 1.

Let's consider the generalized quaternions by looking closely at the second member of the family, Q_{16}. Specifically, the presentation for this group is
$$Q_{16} = < x, y : x^8 = y^4 = 1; x^4 = y^2; xy = yx^{-1} >$$

We expect 16 elements, so we begin by listing the names of the elements we have. The element x generates fully half of the group, and that will always be the case with a generalized quaternion because of the way they are defined. In addition to those 8 elements we have two more elements from y: y and $y^3 = y^{-1}$. The other elements will be products of x and y, so the candidates are xy, x^2y, x^3y, x^5y, x^6y, x^7y, xy^3, x^2y^3, x^3y^3, x^5y^3, x^6y^3, x^7y^3. We need not worry about putting the y first in the product because the third relation in the presentation will automatically determine those products. There are 12 possible new elements and so we expect duplicates here. A little thought shows that duplicates arise from the second relation, $x^4 = y^2$:

$$x^5y = x \cdot x^4y = xy^2 \cdot y = xy^3$$
$$x^6y = x^2 \cdot x^4y = x^2y^2 \cdot y = x^2y^3$$
$$x^7y = x^3 \cdot x^4y = x^3y^2 \cdot y = x^3y^3$$

In other words, if n > 4 we have $x^ny = x^{n-4}y^3$

$$x^5y^3 = x \cdot x^4y^3 = xy^2 \cdot y^3 = xy$$
$$x^6y^3 = x^2 \cdot x^4y3 = x^2y^2 \cdot y^3 = x^2y$$
$$x^7y^3 = x^3 \cdot x^4y^3 = x^3y^2 \cdot y^3 = x^3y$$

and in other words if n < 4 we have $x^ny = x^{n+4}y^3$. Using these duplications we see that we need not use y^3 in naming the product elements, and that we have a total of 16 of them:
$$1, x, x^2, x^3, x^5, x^6, x^7, y, y^3, xy, x^2y, x^3y, x^5y, x^6y, x^7y$$

This does not show that these 16 elements actually do form the group. The next step is to show that the product of any two elements on the list is already on the list; in other words, that multiplication of these elements is well-defined and closed. To do this, consider two arbitrary elements and multiply them, say x^ny and x^my, where both n and m are between 0 and 7. First suppose m < 4. Then

$$(x^ny)(x^my) = (x^ny)(yx^{-m}) \quad \text{(using the third relation m times)}$$
$$= x^ny^2x^{-m} \quad \text{(using the associative law)}$$
$$= x^n \cdot x^4 \cdot x^{-m} \quad \text{(using the second relation)}$$
$$= x^{n-m+4}$$

The last exponent must be computed mod 8 of course. In short, the list of 16 elements we gave is a closed system under the multiplication governed by the given relations.

We will now determine the lattice diagram for Q_{16}, and it will not require much work. We know from the presentation that x generates a copy of Z_8, and that y generates a copy of Z_4 which share its Z_2 subgroup with the Z_8. Our first step, as usual, is to determine what the other six elements generate:

$(xy)^2 = xyxy = yx^{-1} \cdot xy = y^2 = x^4$
$(xy)^3 = (xy)^2(xy) = x^4 \cdot xy = x^5 y$
$(xy)^4 = (xy)^2(xy)^2 = x^4 \cdot x^4 = x^8 = 1$

Thus we have another copy of Z_4 sharing the single copy of Z_2. There remain 4 more elements to consider.

$(x^2 y)^2 = x^2 y \cdot x^2 y = x^2 y \cdot y x^{-2} = x^2 y^2 x^{-2} = x^2 \cdot x^4 \cdot x^{-2} = x^4$
$(x^2 y)^3 = (x^2 y)^2 (x^2 y) = x^4 \cdot x^2 y = x^6 y$
$(x^2 y)^4 = (x^2 y)^2 (x^2 y)^2 = x^4 \cdot x^4 = x^8 = 1$

Another copy of Z_4 sharing the single copy of Z_2. Finally we have

$(x^3 y)^2 = x^3 y \cdot x^3 y = x^3 y \cdot y x^{-3} = x^3 y^2 x^{-3} = x^3 \cdot x^4 \cdot x^{-3} = x^4$
$(x^3 y)^3 = (x^3 y)^2 (x^3 y) = x^4 \cdot x^3 y = x^7 y$
$(x^3 y)^4 = (x^3 y)^2 (x^3 y)^2 = x^4 \cdot x^4 = x^8 = 1$

A final and fifth copy of Z_4 sharing the single copy of Z_2. These are all the cyclic subgroups. We need not look for dihedral subgroups since there is only one element of order 2.

Are we done then? One thing further to check: are there quaternion subgroups Q_8 present? This is very much like checking for copies of V. We are looking for generators of Z_4's that are closed with respect to multiplication. There are five copies of Z_4:

$\langle y \rangle = \langle y^3 \rangle \quad \langle x^2 \rangle = \langle x^6 \rangle \quad \langle xy \rangle = \langle x^5 y \rangle \quad \langle x^2 y \rangle = \langle x^6 y \rangle \quad \langle x^3 y \rangle = \langle x^7 y \rangle$

A superficial examination of these should suggest to you that putting y, x^2, and $x^2 y$ together will generate a copy of Q_8. Further, it is easy to see that putting x^2, xy, and $x^3 y$ together will also generate a copy of Q_8. The calculations are quite easy and are left for you to check. The resulting lattice diagram is shown on the next page.

It turns out that $Q_{2^{\wedge}(n+1)}$ will always contain exactly one copy of $Z_{2^{\wedge}n}$ and two copies of $Q_{2^{\wedge}n}$, as well as exactly one copy of Z_2. You will be asked to prove this in the exercises.

Exercise 89: In the presentation of the generalized quaternions, set n = 1 and determine the resulting group.

Exercise 90: Create a complete Cayley table for Q_{16}.

Exercise 91: Construct the lattice diagram for Q_{32} as completely as possible. Be sure to look for copies of Q_8 and Q_{16} among the subgroups.

Exercise 92: Show that $Q_{2^{\wedge}(n+1)}$ always contains exactly one copy of $Z_{2^{\wedge}n}$ and two copies of $Q_{2^{\wedge}n}$.

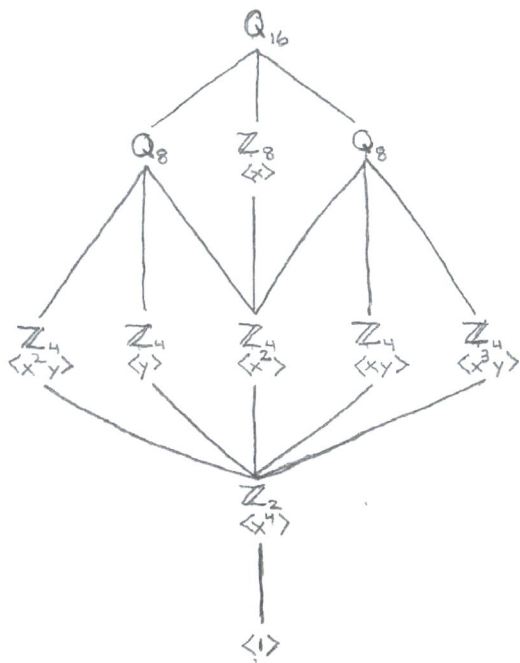

Lesson 31: The Cayley Digraph

For our final lesson introducing groups we will introduce one more way of diagramming the group structure. The lattice diagram shows the structure of the group in terms of its subgroups. The diagram we will introduce now pictures the structure of the group in terms of its elements; specifically in terms of its generators. This gives a more algebraic view of how the group behaves.

To begin we must choose a presentation for the group, but really all we need to choose for this are the generators. A different choice of generators for the group will give a different picture of the group. Next we put dots, which we will call *vertices*, in some pattern on the paper, one vertex for each element of the group, and label the vertices with the elements they represent. Now choose one of the generators, say x. Draw an arrow from each vertex, say the one labeled y, to the vertex which is the element equal to xy. Since we will be making these digraphs for non-Abelian groups we must be careful to always perform the multiplication from the left or always from the right. Here we will adopt the convention that we always multiply the generator on the left to produce the arrow.

Clearly there will be an arrow from the identity element to x. Once we have all the arrows in place, we have a map of how the generator x acts on each of the other elements. There will be a cycle, which will be all the elements in < x >. Disjoint from that cycle there will be other cycles representing how x acts on the elements other than the ones it specifically generates.

Now we choose another of the generators, say z, for the group and do exactly the same thing, drawing arrows from an element y to the element which equals the product zy. Draw these arrows in a different color, however, so the action of x is clearly separated from the action of z. The arrows for this generator will also partition the vertices into disjoint cycles but now you will find that arrows cross between the cycles created by the action of the first generator.

Continue with this process until you have gone through all the generators, using a distinct color for each one. The more generators, the more complicated the diagram will be, of course. Once all of the generators have been represented by arrows, the result is called the **Cayley digraph** of the group. The prefix "di-" in the name is short for "directed". The arrows naturally indicate a direction. If we inserted line segments rather than arrows, we would call the result the **Cayley graph** of the group.

As you might expect, the readability, the usefulness, of the Cayley digraph will depend on where we placed the vertices of the various elements. If the vertices are positioned so that there is a lot of crossing of arrows, the digraph will be hard to make sense of. Therefore the final step of the construction should be to move the vertices to new locations which will untangle the arrows as much as possible. In the end we should be able to arrange the vertices to give a digraph that displays the symmetry of the group operation itself.

To show how this might work out in practice I have shown the Cayley digraph of Q_8 using the generators $\check{k}$, $\check{j}$, and $\check{i}$. For the first digraph I simply arranged the vertices in what seemed to be a logic formation. In the right hand digraph, the vertices have been rearrange to produce a much more pleasing digraph, pleasing both aesthetically and practically. The utility of our algebraic tools depend critically on how careful we are with the aesthetics, whether we are making a lattice diagram or a Cayley digraph. Ugly diagrams almost always cause mistakes. In the digraphs of Q_8 I used black arrows to represent multiplication on the left by $\check{j}$, and red arrows to represent multiplication on the left by $\check{i}$.

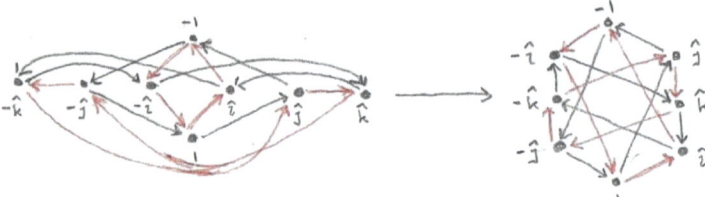

If we change the presentation we are using, then the Cayley digraph will also change. This can be very help in giving a different perspective on how the group operation works. The dihedral groups, for example, have two very different presentations. In one we use the basic rotation and a reflection to generate the group and in the other we use two reflections to do it. Below I have given both Cayley digraphs for you to compare. In the left hand digraph the black arrows indicate left multiplication by ρ_1 and the red arrows indicate left multiplication by δ_1. In the right diagram the black arrows represent left multiplication by μ_2 and the red arrows represent left multiplication by δ_1. I have given only the neatened digraph and not the tangled one that it came from.

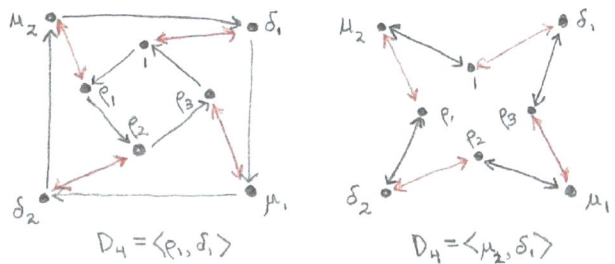

One easy use of the Cayley digraph is that, by simply following arrows, we can write any element as the product of generators multiplying on the left beginning with any other element. For example, in $\mathbf{D}_4$ generated by the two reflections, suppose I wanted to express ρ_3 as a product of a string of the two generators multiplying δ_2. Then I would start at δ_2 and choose a path that ends at ρ_3. Note that all the arrows are pointed in both directions, so I need not be careful about that. It should be easy for you to check that $\rho_3 = \delta_1 \cdot \mu_2 \cdot \delta_1 \cdot \delta_2$. Of course, $\mathbf{D}_4$ is simple enough that we do not need such tricks to help us do any calculation. For more complicated or unfamiliar groups, however, the Cayley digraph can be quite useful for expressing such products as well as giving us some insight into the structure of the group. This is only one use of these digraphs, and we can't make full use of them at this point. Here they are just presented as one more way of picturing the structure of a group, a viewpoint to supplement what we already have to reveal perhaps unexpected aspects of the algebra.

On the next page I will list all the groups we have studied in this book, arranged by their order and classified as either Abelian or non-Abelian.

Exercise 93: Construct the Cayley digraph for $\mathbf{Q}_{16}$.

Exercise 94: Construct two Cayley diagraphs for $\mathbf{D}_6$.

Exercise 95: Assuming $\mathbf{A}_4$ is generated by the two 3-cycles $(1, 2, 3)$ and $(1, 2, 4)$, construct the Cayley digraph which proves it is.

Table of Groups to Order 32

order	Abelian	non-Abelian	
2	Z_2		
3	Z_3		
4	Z_4, V		
5	Z_5		
6	Z_6	$S_3 = D_3$	
7	Z_7		
8	Z_8, U_{15}	D_4, Q_8	(1 missing)
9	Z_9		(1 missing)
10	Z_{10}	D_5	
11	Z_{11}		
12	Z_{12}, U_{21}	D_6, A_4	(1 missing)
13	Z_{13}		
14	Z_{14}	D_7	
15	Z_{15}		
16	Z_{16}	D_8, Q_{16}	(11 missing)
17	Z_{17}		
18	Z_{18}	D_9	(3 missing)
19	Z_{19}		
20	Z_{20}	D_{10}	(3 missing)
21	Z_{21}		(1 missing)
22	Z_{22}	D_{11}	
23	Z_{23}		
24	Z_{24}	D_{12}, S_4	(12 missing)
25	Z_{25}		(1 missing)
26	Z_{26}	D_{13}	
27	Z_{27}		(4 missing)
28	Z_{28}	D_{14}	(2 missing)
29	Z_{29}		
30	Z_{30}	D_{15}	(2 missing)
31	Z_{31}		
32	Z_{32}	D_{16}, Q_{32}	(48 missing)

Index of Terms

Abelian group	9
alternating groups	22, 23
associative law	2
associative operation	2
bijection	4
bijective function	4
cancellation laws	6
Cayley digraph	31
Cayley table	11, 12
commutative law	9
commutative operation	9
complex numbers	1
composition	2
congruence equation	11, 14, 18
cube symmetries	27
cycle	20
cycle notation	20
cyclic groups	9, 10, 14, 16. 17, 18
dihedral groups	24, 25, 26, 28
disjoint cycles	20
Euclidean algorithm	13
Euler's Phi function	15
Euler's Theorem	15
even permutation	21, 22
fix a letter	19, 23
fraction	5
generalized quaternions	30
generator	8, 9, 28
greatest common divisor	9, 11, 13
group	5
group of units	11, 12, 13, 14, 15, 16, 17, 18
identity	3
improper subgroup	7
index (of an element in U_n)	18
injection	4
injective function	4
integers	1
intersections	7, 9
inverse	4, 5, 13
join	7, 9
Klein 4-group	11, 25, 28
lattice diagram	10, 11, 12, 22, 23, 24, 25, 28
least common multiple	9
left identity	3
left inverse	4

length (of a cycle)	20
letter	19
modular law	10
monoid	3, 4, 11
move a letter	19
n-cycle	20
natural numbers	1
odd permutation	21
operation on the set S	1
order of a group	7, 9
order of an element	8, 9
permutation	19
permutation notation	19
power set	1, 2
presentation	28, 29, 30
primitive root	16, 17
proper subgroup	7
symmetry	24, 25
quaternion group	29
rational numbers	1
real numbers	1
relation	28
right identity	3
right inverse	4
semi-group	2, 3
stabilizer	23
subgroup	7
subgroup generated by g	8, 9
subtraction	5
surjection	4
surjective function	4
symmetric group	19, 20, 23, 26
transposition	20, 21
trivial subgroup	7
two-sided identity	3
two-sided inverse	4
unit	11

Index of Symbols

Symbol	Page
A_n	22
$\mathbb{C}$	1
$\mathbb{C}^*$	1
D_n	24
$\mathbb{E}$	1
$F(S)$	2
$\langle g \rangle$	8
$\mathbb{I}$	1
$\mathbb{I}^0$	1
ind_r	18
$\mathbb{N}$	1
$\mathbb{O}$	1
O^*_{48}	27
$P(S)$	1
$\mathbb{Q}$	1
$\mathbb{Q}^+$	1
$\mathbb{Q}^*$	1
Q_{2^n}	30
$\mathbb{R}$	1
$\mathbb{R}^+$	1
$\mathbb{R}^*$	1
S_n	20
$Stab(x)$	23
U_n	11
V	11
$\mathbb{Z}$	1
Z_n	9
$\varphi(n)$	15
$\leq$	7
$\forall$	2
$\in$	2
$\vee$	7

www.ingramcontent.com/pod-product-compliance
Lightning Source LLC
Chambersburg PA
CBHW051157220526
45473CB00003B/803